S0-BCM-185

August-Wilhelm Scheer

CIM
Computer Integrated
Manufacturing

Towards the Factory
of the Future

Third, Revised and Enlarged Edition

With 155 Figures

Springer-Verlag

Berlin Heidelberg New York
London Paris Tokyo
Hong Kong Barcelona
Budapest

Professor Dr. August-Wilhelm Scheer
Universität des Saarlandes
Institut für Wirtschaftsinformatik
Postfach 15 11 50
D-66041 Saarbrücken, FRG

Library of Congress Cataloging-in-Publication Data

Scheer, August-Wilhelm.
   [CIM. English]
   CIM : computer integrated manufacturing : towards the factory of
the future / August-Wilhelm Scheer. -- 3rd, rev. and enl. ed.
      p.   cm.
   Includes bibliographical references and index.
   ISBN 0-387-57964-8 (acid-free paper : U.S.)
   1. Computer integrated manufacturing systems. 2. Computer
integrated manufacturing systems--Germany.   I. Title.
TS155.63.S34   1994
670'.285--dc20   94-31567
                                                          CIP

ISBN 3-540-57964-8 Springer-Verlag Berlin Heidelberg New York Tokyo
ISBN 0-387-57964-8 Springer-Verlag New York Heidelberg Berlin Tokyo
ISBN 3-540-53667-1 Springer-Verlag Berlin Heidelberg New York Tokyo
ISBN 0-387-53667-1 Springer-Verlag New York Berlin Heidelberg Tokyo

This work is subject to copyright. All rights are reserved, whether the whole or part of the material is concerned, specifically the rights of translation, reprinting, reuse of illustrations, recitation, broadcasting, reproduction on microfilms or in other ways, and storage in data banks. Duplication of this publication or parts thereof is only permitted under the provisions of the German Copyright Law of September 9, 1965, in its version of June 24, 1985, and a copyright fee must always be paid. Violations fall under the prosecution act of the German Copyright Law.

© Springer-Verlag Berlin · Heidelberg 1988, 1991, 1994
Printed in Germany

The use of registered names, trademarks, etc. in this publication does not imply, even in the absence of a specific statement, that such names are exempt from the relevant protective laws and regulations and therefore free for general use.

2142/2202-543210

Preface to the 3rd Edition

As against the 2nd edition especially Part D „CIM Implementations" has been revised. It is supplemented by articles on „HP OpenCAM"—the new CAM strategy of Hewlett Packard GmbH—and the reengineering of business processes. The articles are based on the Architecture of Integrated Information Systems (ARIS) developed by the author.

Part E „CIM Promotion Measures" is extended by articles on CIM in Eastern Europe, Brazil and China. The article on CIM-Technology-Transfer-Centers has been revised.

I would like to thank all the named contributors for the punctual preparation and delivery of their manuscripts.

I would also like to thank Mr. Wolfgang Hoffmann for his dedicated support in the technical preparation of the manuscript.

Saarbrücken, Germany

June 1994

August-Wilhelm Scheer

Preface to the 2nd Edition

The fears of the author, expressed in the preface to the 1st edition, that, although the CIM philosophy is correct, its implementation would be frustrated by the EDP-technical difficulties, have fortunately proved unfounded. Instead an increasing number of successful CIM implementations are appearing. The experience gained in the interim in implementing CIM is extensively incorporated in this edition. The "Y-CIM Information Management" model developed at the Institut für Wirtschaftsinformatik (IWi) is presented as a tried and tested procedural approach to implementing CIM.

The examples presented have also been brought up-to-date.

Presentation of the CIM Center set up by manufacturers has been omitted, however, since examples of concrete applications have in the meantime replaced the interest in "laboratory versions".

In addition to German examples, the experience of American CIM users is also presented. The primary aim is to allow comparative assessments. But, at the same time, this serves to demonstrate the way that American EDP manufacturers are thinking, which will have a decisive influence on the development of hardware and software products for CIM.

The sections in the previous edition which provided surveys of further CIM developments: design stage cost estimation, the use of decision support systems (expert systems), and inter-company process chains, have in the meantime proved themselves to be effective CIM components and are therefore incorporated in the general text.

I would like to thank all the named contributors for the punctual preparation and delivery of their manuscripts.

I would also like to thank Ms. Irene Cameron for her careful translation of the German original, and my employees, particularly Mr. Carsten Simon, Ms. Rita Landry-Schimmelpfennig, and Ms. Carmen Kächler, for their dedicated support in the technical preparation of the manuscript.

Saarbrücken, West Germany   August-Wilhelm Scheer
December 1990

From the preface to the 1st edition

This book appeared in the Federal Republic of Germany in 1987, and within one year it had run to 3 editions. This indicates the strength of interest German industry is showing in CIM integration principles. CIM is, however, a concept of international relevance to the structuring of industrial enterprises. The author has made several research visits to CIM development centers of leading computer manufacturers and important industrial enterprises which have indicated that the stage of development of CIM in Germany in no way lies behind the international level. Rather, in the USA for example, considerable uncertainty exists regarding the status of manufacturing. For this reason, the CIM examples presented in this book are also of interest to US industry.

The stance taken in this book of defining CIM as a total concept for industrial enterprises is increasingly shared internationally. CIM is more than CAD/CAM.

Although CIM has been broadly accepted both in theory and in practice, its development is nevertheless subject to risks. A concept which is in itself correct can still come to grief if the implementation possibilities are not yet adequateley developed. So, in the 1960s, the MIS concept failed because the necessary database systems, user-friendly query languages, and comprehensive, operational base systems for providing data were not available.

CIM must learn from this experience. Hence, it is important to convince the interested user as quickly as possible that the implementation of CIM is possible with currently available computer technology. But not only suitable computer tools are needed for the implementation of CIM, equally important is the new organizational know-how: a transition from specialized, subdivided operational processes to integrated, unified processes is needed.

Given the economic interests of computer manufacturers, great efforts are currently being made to develop new hardware and software concepts for CIM. Here too, therefore, favourable prerequisites for the implementation of CIM are increasingly being met. A significant bottleneck, however, is training and retraining in integrated CIM concepts.

This book, therefore, does not aim to put the functional details of the individual CIM components (PPC, CAD, CAP, and CAM) in the foreground, but rather to emphasize the integration principles and elaborate the implications of the integration principles for the functional demands of the individual components.

Saarbrücken, West Germany

January 1988   August-Wilhelm Scheer

Contents

Introduction

In coming years the introduction of Computer Integrated Manufacturing (CIM) will become a matter of survival for many industrial concerns. Information technology will increasingly be recognized as a factor of production, not only influencing organizational structure, but also becoming a significant competitive factor.

The resulting link between information technology and organizational procedures will be exploited not only by large corporations; even in middle to small scale enterprises it will become an important factor in corporate policy. This will also occur because increasing inter-company cooperation will spread CIM principles from the larger to the smaller enterprises.

The important factors motivating the introduction of CIM are the cost advantages and the flexibility it can yield. These cost advantages, generated by integration and streamlining of processes, need to be exploited in the current climate of increasing international competition. Increased flexibility within the production process is necessitated by a more client-oriented environment with shorter product life-cycles and the correspondingly higher need for replacement parts.

This book addresses itself to the following questions:

- What are the components of Computer Integrated Manufacturing?
- What are the data relationships between the components?
- How can a strategy for introducing a Computer Integrated Manufacturing system be developed?
- What CIM prototypes already exist?
- What are the potential future developments from CIM?

A. The Meaning of the "I" in CIM

Computer Integrated Manufacturing (CIM) refers to the integrated information processing requirements for the technical and operational tasks of an industrial enterprise. The operational tasks can be referred to as the production planning and control system (PPC), as represented in the left fork of the Y in Fig. A.01. The more technical activities are

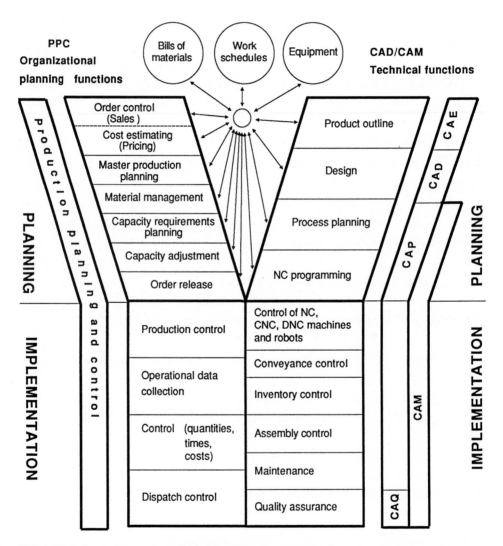

Fig. A.01: Information systems in production

characterized by the various CAX-concepts in the right fork of the Y. The PPC system is determined by order handling, whereas the CA-components support product description and the production resources. At the same time these information systems provide data for the associated financial and cost accounting systems.

The integration of these areas makes particularly high demands on the willingness of enterprises to face up to the integration requirements at the organizational level. It is also a challenge to hardware and software producers finally to coordinate their separately developed systems for technical and business use.

I. Data and Operations Integration

In this century Taylorism, with its functional division of responsibility, has been a dominating guideline for the structuring and running of an organization. This is depicted in Fig. A.I.01,a in which an essentially unified process is divided into three sub-processes, each carried out by separate departments. In each sub-process a lead-in period occurs, and each department manages its own data. Information about the

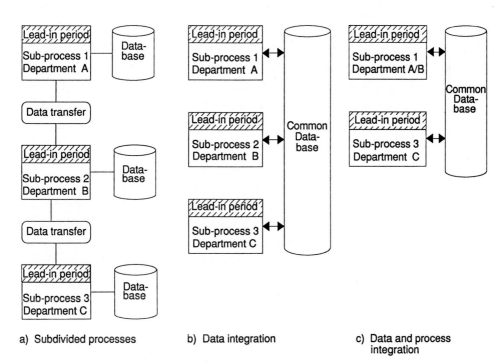

a) Subdivided processes   b) Data integration   c) Data and process integration

Fig. A.I.01: Reintegration of functionally divided operations

processing stage reached in the previous sub-processes must, therefore, be transferred in all its complexity between the individual departments.

Given the strong specialization on which Taylorism is based, benefits may result from accelerated handling within the **sub-processes**. Nevertheless, many empirical production and management studies have shown that, where processes are sub-divided and specialized, throughput times are extremely high as a result of the repeated lead-in and data transfer times. More specifically, communication and lead-in times of the order of 70% - 90% for administrative order handling and production processes have been recorded. This high share represents a considerable potential for rationalization, since long throughput times lead to high capital tie-up. This can involve substantial competitive risks in an age of just-in-time production with its increased demand for consumer-oriented flexibility.

How can CIM affect this situation?

One reason for the cumbersome information transfer represented in Fig. A.I.01,a is the specialized, departmentally determined data organization. In Fig. A.I.01,b the entire process uses one common database, which allows information accruing at one stage in the production chain to be included in the database, and thus to become immediately available to all other stages in the chain. In this way information transfer times are eliminated, and processes can be considerably accelerated. In recent years this principle of integrated data processing has already been largely achieved, and has led to rationalization **within** divisions of the enterprise, such as accounts, production planning and control and order handling. For example it has been possible to reduce administrative order handling times from 3 weeks to 3 days.

With regard to CIM, the realization of this integration principle necessitates the creation of the relevant data links between the technical areas of design, process planning and production and the corresponding administrative processes, such as production planning and control. This means that information systems which are **in themselves** already partly integrated must now be unified **with each other**, due to the fact that technical and organizational activities involve increasing interaction within the processing chain of a given customer order.

The rationalization of the integrated data processing system for production can occur only when the control loop is closed. Up-to-the-second online processing achieved in one sub-division can be rendered incapable of altering the total

throughput time of the process chain if it is dependent on a batch process data transfer occurring only daily or weekly.

A further implication of the CIM model results from the fact that sub-functions within the operation chain can be more powerfully reintegrated; i.e. excessive specialization can be eliminated.

An important reason for the establishment of specialized processes was that the human capacity for information processing is strictly limited, and therefore only subsets of the total process could be viewed and handled. With the support of database systems and user-friendly interactive processing systems, the human ability to handle complex tasks has grown. As a result the reasons which earlier forced strict specialization **no longer hold**, and sub-processes can be reintegrated at the workplace.

This is represented on the left-hand side of Fig. A.I.01,c where sub-processes 1 and 2 are combined. Consequently, lead-in times occur only once, and data transfer times are completely eliminated between sub-processes 1 and 2.

Both these effects, data integration and process integration at the workplace, jointly give rise to the high rationalization potential of CIM.

II. A Typical CIM Process Chain

This general principle of integration reflects the interdependence of organizational and technical processing functions. Fig. A.II.01 depicts a typical order handling operation, organized along specialized lines, as a process chain.

In each department computer systems are already in use, but the flow of information **between** departments takes place on paper. Data from the order receipt system are sent on a paper form to the CAD system in the design department. The same is true of the data flow between the design and process planning departments, in that the drawing is used as the basis for work scheduling, and hence important information already included in the drawing must be manually re-entered into the computer-based information system.

6

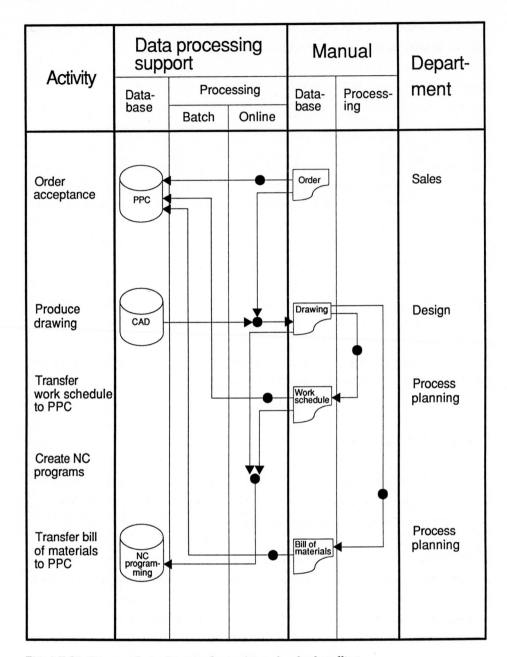

Fig. A.II.01: Process chain diagram for traditional order handling

The transfer of data from design to production also gives rise to complicated transfer procedures and hence lost time. Geometry data already included in the CAD system and required for NC programming must be read from the drawing and re-entered. Information needed for production planning and control, such as bills of materials, although already clearly specified in the design department, must also be entered once more into the primary data of the PPC system.

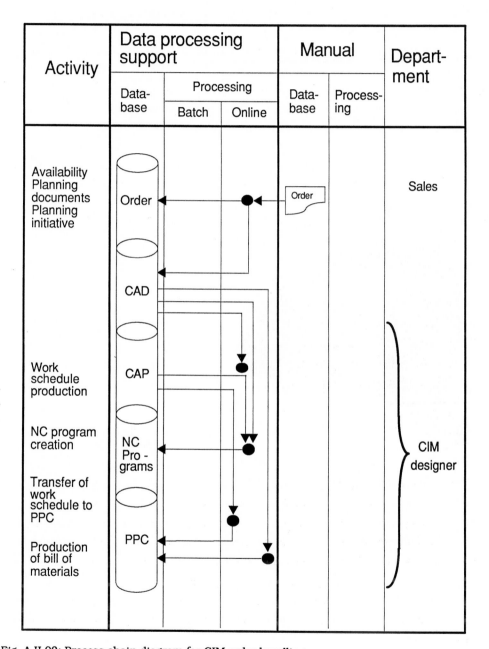

Fig. A.II.02: Process chain diagram for CIM order handling

Fig. A.II.02 depicts the equivalent CIM integrated operation, which includes both **data** and **process integration**:

Customer requests for a particular product variant are recorded by the order receipt department, and immediately transferred via the common database to the design department. Using a similarity schedule based on previously designed and manufactured products the design department can then estimate the complexity of the customer request in both production and cost terms. If only minimal alterations are required information from drawings in the CAD system can be forwarded to the customer. Including the drawing with the offer can favor acquisition. Once the order has been taken CAD can carry out the detailed design and specify the geometry exactly.

In Fig. A.II.03 a ball-bearing is shown as a 3-D model in the form of an explosion drawing. This illustration can be used to demonstrate further the data flows within the chain.

Fig. A.II.03: Ball-bearing
Source: *IBM*

The explosion drawing clearly depicts the construction of the ball-bearing. It consists of several rings and a specified number of balls. Hence the bill of materials of this ball-bearing, as presented in Fig. A.II.04, can be derived directly from the geometric data. The concept of integrated data management therefore requires that the information implicit in the CAD system can be assimilated into the primary data for the bill of materials of a PPC system. Only through integrated data management is it possible to maintain the consistency of large volumes of data-quantities which already exist in connection with the primary data management of a PPC system, and which will quickly accumulate in the CAD area in coming years.

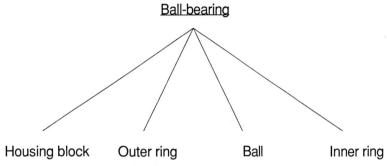

Fig. A.II.04: Bill of materials for a ball-bearing

At the same time, the geometric representation of the ball-bearing also contains information which can be further utilized in production. For example, the position and diameter of the holes to be bored are also given in the drawing. Thus it is already possible to check whether the appropriate bit for the NC drilling system is available or not, so that, if need be, the drawing can be modified to a design suitable for production. In this context direct access from the CAD system to the machine and tool database is necessary. Once this check has been successfully carried out, the information as to the position and diameter of the drill holes can be transferred to the NC control program, and the NC programmer merely needs to supply the technological data, such as drilling speed. The direct transfer of geometry data from the CAD system to the NC programming system in CAM has now become a recognized requirement. Its implementation, however, continues to meet with problems.

Work scheduling also accesses the engineering drawing, and can be largely automated as a result of its close relationship with NC programming by using computer-managed tables. Only raw material dimensions and machining parameters need to be entered.

The CIM process illustrated in Fig. A.II.02 demonstrates that, in comparison with the unintegrated process of Fig. A.II.01, **all** interfaces are effected via a unified database, and paper forms become obsolete. At the same time the marketing, design and process planning systems are more closely connected. The term **CIM design** as a departmental name expresses the intense unification of design and production planning that follows from CIM principles. The CIM designer is also involved in order processing; where the customer deadline is tight he can check the availability of the necessary materials, and hence function as a material requirements planner. The demand for design which is suitable for manufacture requires that the functions of process planning and, in the course of design-stage cost estimation, accountancy also be taken over.

The fact must also be recognized that in the areas of design and product development the introduction of CIM gives rise to a new decision-making center within the industrial concern: Here material requirements are established, the choice between in-house production and external purchasing influenced, and, on account of the closer contact between design and production through increased automation, the production method determined. For these reasons the early incorporation of cost considerations, as expressed in the phrase "**design-stage cost estimation**", is a strict requirement. This requirement is all the more important when it is recognized that the factors affecting cost assessability and cost controllability are diametrically opposed in their relation to the design stage reached (see Fig. A.II.05).

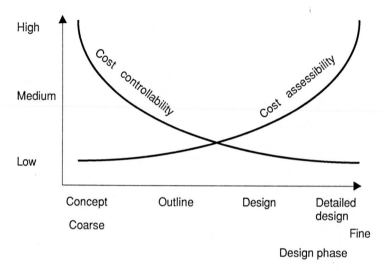

Fig. A.II.05: Cost effects in the design phase

The multi-faceted relationships between PPC, CAD, CAP and CAM, as they occur in practice, could be only vaguely suggested in our example. In general though, it is a fair assertion that the imagination never goes far enough in conceiving of possible data links and their permutations. A few should, however, be mentioned here:

-    The provision of precision-fitting packaging for shipping fragile parts will require geometric data from the CAD system (link between CAD and shipping control).
-    In CAD collision tests can be conducted using simulated production runs in which, for example, the milling process is depicted on the display screen once the contours of a part to be processed have been established. From here it is a small step to estimate production time, and hence the relationship to work scheduling and cost estimation of the process (link between CAD, cost accounting, CAM and CAE).

- Production requires not only information about the part to be processed, i.e. order information, but also manufacturing instructions in the form of NC programs. Before a production order can be given clearance, therefore, it is necessary that availability checks be conducted not only on the required materials, components, labor and production facilities, but also on the NC program library (link between PPC and CAM).

- In the course of operational data collection information about orders, production facilities, inventories and personnel will be recorded and will constitute the basic inputs to the control functions of a PPC system. Here, too, there is a close link between technical and organizational aspects. The increasingly intelligent control of manufacturing equipment is more and more capable of accumulating order information for both counted and weighed processes and directly entering it into the operational data collection system. From there it can be used not only for the compilation of production statistics, but also for certification within a quality assurance system, or for productivity-based wage calculation (link between CAM, data collection, PPC, CAQ and pay-roll accounting).

III. The CIM Enterprise: The Computer Steered Industrial Firm

Strict adherence to the CIM integration principle within an enterprise not only implies the reorganization of individual process chains, but also the integration of all information flows within a unified system. Only then can one talk of a CIM enterprise, of the computer-steered industrial firm.

Fig. A.III.01 presents the CIM model of the Californian glass manufacturer Guardian Industries.

The firm has made personal computers available to its most important customers free of charge. The computers are used to provide the customers with data relating to Guardian Industries' production program, in particular as regards article numbers, descriptions, production dimensions, possible colors and prices. These data are kept up to date for the customer by the firm via a data network.

Customers inform the firm of orders by accessing the stored data and transferring it to a screen form, whereby the plausibility of being able to deliver the desired colors, dimensions, etc. is checked. In this the way the orders that the firm receives already contain the correct order data. This means that for Guardian Industries the usual reconciliation procedures involved in order recording are eliminated.

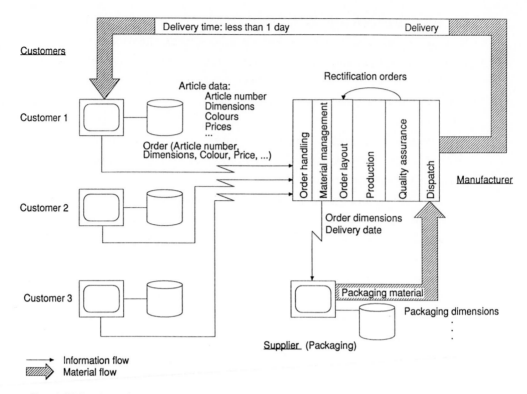

Fig. A.III.01: Guardian Industries' CIM model

The incoming orders are aggregated and distributed on the sheets of raw glass using a layout optimization program, so as to minimize wastage. Once the orders have been distributed to the sheets of raw glass the geometry for the subsequent cutting operations is established. This geometry resulting from the order assignment automatically generates the cutting programs for the cutting robots.

The production process is monitored continually. Production errors lead to the creation of re-orders, so that at the end of the production run complete customer orders have always been produced. The job of managing part orders is therefore unnecessary.

From the readily available order data, supplier orders are automatically communicated via remote data processing to the supplier. On the basis of these data the supplier can produce customized packaging materials.

The interdependence of order recording, production planning, material management, quality assurance and dispatch makes it possible for Guardian Industries to guarantee a delivery time of less than one day: in concrete terms this means that all orders that the firm receives before 4pm can be dispatched by 8am the following morning.

This example demonstrates the holistic nature of CIM: the entire order flow from the customer through production to the suppliers is regarded as a **unified** task. In contrast, in an organization with specialization the sub-systems sales, production planning and control and material management are each regarded as independent tasks, which are merely linked via interfaces where necessary. This is the consequence of an organizational structure based on functional breakdown.

That CIM here constitutes the heart of a unified enterprise strategy is also clear from the written CIM strategy of Guardian Industries, which specifies the following fundamentals:

- Guardian is market cost leader.
- As a result of the 24-hour operation of the computer system orders can be accepted around the clock.
- There are no stocks of semi-finished or finished goods.
- Important customers are bound to the firm by the order acceptance service.
- Important customers are responsible for the accuracy of order data.
- Only complete customer orders are delivered.
- Orders which are received by the firm before 4pm are dispatched by 8am the following day.

Each of these factors makes economic sense, and they give rise to the demands on the computing system as regards:

- reliability of the computing components in maintaining 24-hour operation,
- the use of a relational database system to support data integration,
- realtime operation of process control,
- the use of communication networks to link the sub-systems.

The current discussion of CIM often tends to be couched more in terms of the second set of considerations, that is in instrumental terms, while the fundamental possibilities of the strategic implications of a consistent exploitation of CIM technology remain neglected. However, this discussion makes it clear that CIM is essentially an enterprise strategy problem.

The design and implementation of a unified CIM system is complicated and requires comprehensive organizational, computer-technical and production-technical knowledge. Once the system is established, however, the routine planning and control functions are simplified.

In the example considered above the organizational streamlining of the process eliminates the management of stocks of semi-finished goods, the management and planning of part deliveries, etc. In short, the more unified the organization of process chains the lower the coordination costs within the chain. This is the consequence of reduced specialization, in which the efficient coordination of temporally independent links in the chain occupies the foreground.

Shifting the weight of routine planning and control onto the correct CIM configuration can also be illustrated using other examples. In the creation of IBM's highly automated plant for producing electronic typewriters at Lexington, USA, the fundamental requirement was that the automation of the factory should reduce production costs to 1/3 of their original level. The entire plant was constructed on the basis of this economic requirement. However, continuous intervention in the production process can scarcely alter the costs arising from the investment decisions.

IV. CIM Definitions

The example described has already illustrated the wide conceptual definition of CIM implicit in the term "computer steered industrial firm" presented here. The concept is increasingly being adopted both in theory and practice.

In Harrington's book "Computer Integrated Manufacturing" which appeared in 1973, and from which the term CIM derives, in addition to CAD and CAM, PPC was also introduced as a component of CIM. The essential treatment of the book, however, was restricted to the level of the manufacturing process (CAM). As as result, until the start of the 1980s there dominated in the USA and Japan a definition of CIM narrowly related to manufacturing and product development in which CIM = CAD + CAM.

However, the comprehensive integration concept of CIM was elucidated early on by computer manufacturers - at least in the form of transparencies for overhead projectors. Fig. A.IV.01 demonstrates this using a diagram originally created by General Electric at the start of the 1980s. In the center of the diagram is a common database accessible by individual users, such as design, production planning and control, parts production, assembly and inventory control. Inter-company cooperation is also possible via external network links.

CIM involves the following:
- an application-independent data organization,
- consistent process chains,
- small feedback loops.

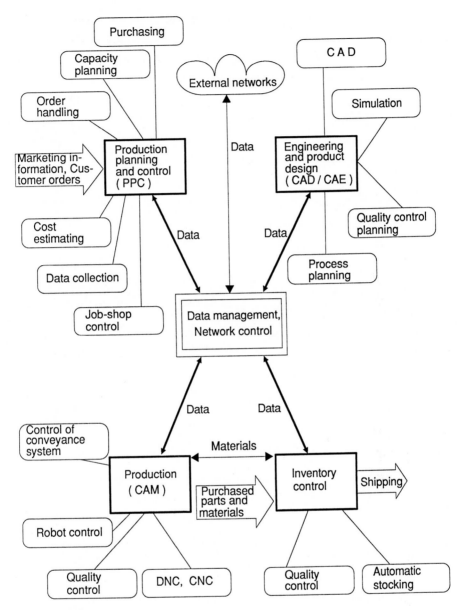

Fig. A.IV.01: Functional integration on the Factory of the Future
   after: *General Electric*

An **application-independent data organization** means that data structures are designed independent of their individual applications. They should be specified so generally that they are available for various tasks. Nowadays this requirement is normally imposed on the design of database-oriented information systems. In concrete terms this means, for example, that there will be only one product specification within the enterprise, which will then be available to technical (product design), production planning (material requirements management) and financial users (product pricing). This is expressed in Fig. A.IV.02 which shows how use of a common database connects organizational and technical aspects in one interlocking process.

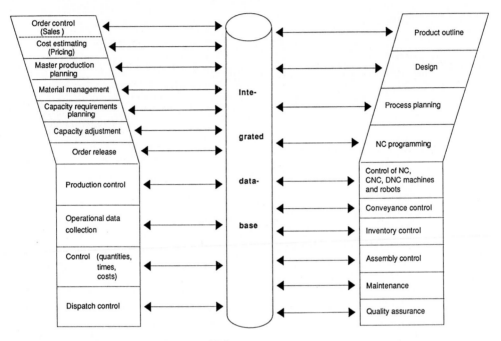

Fig. A.IV.02: Integrated database in CIM

In addition to data integration, as achieved through an application-independent data organization, the concept of **process chains** is characteristic of CIM. Processes are considered in terms of their connectedness, independently of natural organizational structures, and are supported by closed information systems. The notion of **small feedback loops** means that, within any process, planned-actual comparisons are made continually so that, in the event of deviation the control process can intervene quickly. This requires consistent, immediate information processing and a certain decentralization of control responsibility for short-term corrective measures.

Alongside the integration of production planning and control with the more technically-oriented data processing functions, greater integration of commercial activities is also being discussed within the CIM framework. Fig. A.IV.03 depicts this by adding the function CAO (Computer Aided Office) to the other CIM components. This extends the Y-model into an X-model.

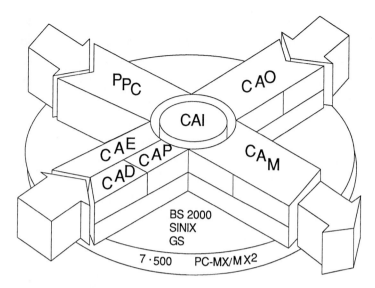

Fig. A.IV.03: Computer Aided Industry (CAI)
            Source: *Siemens*

This extension is of course obvious, but thereby perhaps superfluous. Since there is no such thing as application-independent office work, but rather office jobs which, in an industrial concern, are always associated with a certain application (e.g. product design, process planning, purchasing, cost accounting), a CIM concept which incorporates all the essential tasks of an enterprise, automatically includes the office tasks, too. It is also obvious from the development of information systems to date that the managerial functions of financial and cost accounting are increasingly taken care of in prior functions of the operation. For example, accounts receivable will largely be supplied with data from the invoicing phase of the order processing system, and accounts payable from the purchasing system of material requirements management. Cost accounting obtains current actual data from the operational data collection system, and requires primary data from bills of materials and work schedules to make the necessary product pricing calculations.

Thus, any restructuring of production planning and control in the context of a CIM system automatically relates to managerial system functions, too. Fig. A.IV.03 simply represents a further clarification of this connectivity.

The concept of CIB (Computer Integrated Business) or, better expressed, CIE (Computer Integrated Enterprise) also stress the holistic nature, but do not in any way extend the concepts defined and presented here.

The *Ausschuss für Wirtschaftliche Fertigung e.V.* (*AWF* - Committee for Economic Production), Eschborn, has developed a conceptual representation on the basis of the Y-model which has been widely cited in the literature (see Fig. A.IV.04). It differs from the Y-diagram of Fig. A.01 principally in that it regards quality assurance as an activity accompanying the entire production process. This emphasis derives from the huge importance of quality assurance, particularly for automated production processes. This view is not in conflict with that presented in Fig. A.01, since there the functions belonging to CAM are not ordered in any logical progression, but simply listed as CAM components. For this reason no importance should be attached to the ordering. Furthermore, it should also be noted that quality assurance issues arise not only in the production sphere, but although product development and in the entire product flow, from purchase planning and receiving through output control and shipping.

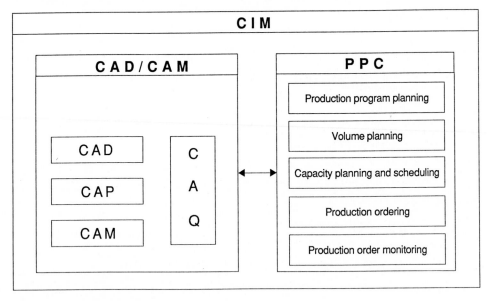

Fig. A.IV.04: AWF recommendation "CIM"
from: *Hackstein, CIM-Begriffe sind verwirrende Schlagwörter 1985, p. 11*

B. The Components of CIM

The CIM components represented in the Y-diagram of Fig. A.01 are:

- Production Planning and Control,
- Computer Aided Design,
- Computer Aided Planning,
- Computer Aided Manufacturing,
- Computer Aided Quality Assurance,
- Maintenance.

A brief description of these components will follow, in which the content of the individual functions will be critically assessed as a basis for the subsequent integration discussion.

I. Stage of Development of CIM Components

a. Production Planning and Control

Production Planning and Control (PPC) is a classic field of application for electronic data processing. Although over the past 20 years many industrial enterprises have dedicated considerable resources to its introduction, the current level of application is not entirely satisfactory. One cause of this is that many enterprises have been overstretched by the introduction of such a complex system, and hence have got "stuck" half way. The complexity arises from the fact that production planning and control accompanies the entire production process (see the functions represented in the left fork of the Y in Fig. A.01).

At the same time rapid hardware and software developments have caused an equally rapid aging of these expensively implemented developments, as regards their functionality and user-friendliness, leading to high reorganization and software maintenance costs.

Suppliers of PPC systems as well as advanced in-house developments of large industrial concerns have established a far-reaching planning concept which builds on the idea of **successive planning**. Here individual planning levels, built on each other, are carried out in logical and chronological order. These planning levels are accompanied by a unified primary data management system.

1. Primary Data Management

Primary data management within a computerized production planning and control system generates the source data necessary for the planning of material and capacity management. At the same time it yields the data needed for the production plan for a specific production order, which is the basis of production control. The production plan contains the essential information needed for production (see Fig. B.I.01).

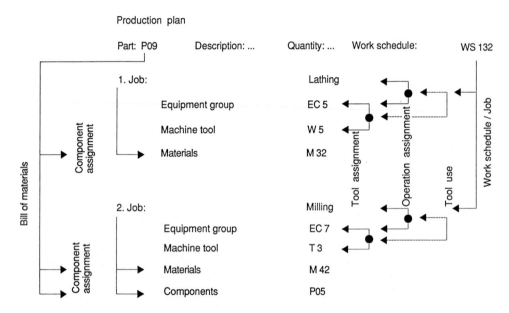

Fig. B.I.01: Content of a production plan

Given the large volume of data particular organizational forms have been established which minimize redundant data storage. This may apply to the construction of a part from its components (bill of materials), production instructions (work schedule and operations), the equipment to be employed (equipment groups), the tools required and the relations between them. Fig. B.I.01 shows this informational decomposition of the production plan.

The construction of parts from their components can be represented diagrammatically by means of a **gozintograph** (see Fig. B.I.02). The gozintograph shows which lower level parts are used in what quantities to construct a given higher level part.

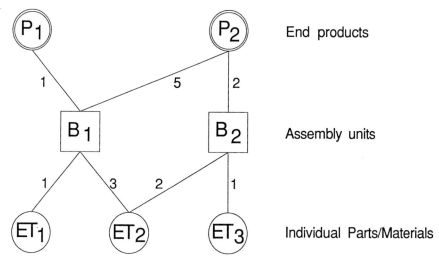

Fig. B.I.02 Gozintograph

Fig. B.I.03 depicts the data structure of a bill of materials with the help of a Chen **Entity Relationship Model (ERM)** (see *Scheer, Principles of efficient information management 1991; Scheer, Enterprise-Wide Data Modelling 1989, p. 17 ff.*). Each object type (entity type) is represented by a box and each relation between entity types by a rhombus; the number of attributes of a relationship that can arise from the entity side is also indicated. In the representation of a bill of materials the entity type will be the number of parts, and the structural relationships will form a relationship type of the form n:m. This indicates that a part can be used to construct several higher level parts, and can itself be constructed from several lower level parts. Other data relationships are also represented in Fig. B.I.03.

The production instructions are incorporated in the work schedule. A part can be produced using various production methods, depending, for example, on the desired degree of automation of production given the quantity to be produced. Conversely, several different parts can be produced using the same (standard) work schedule. A work schedule consists of several operations in which the individual technical procedures for the production of the part are described. A specific operation can also be incorporated in several work schedules. Since the lower level components and materials used to produce a higher level part can be incorporated into the higher level part in a variety of operations

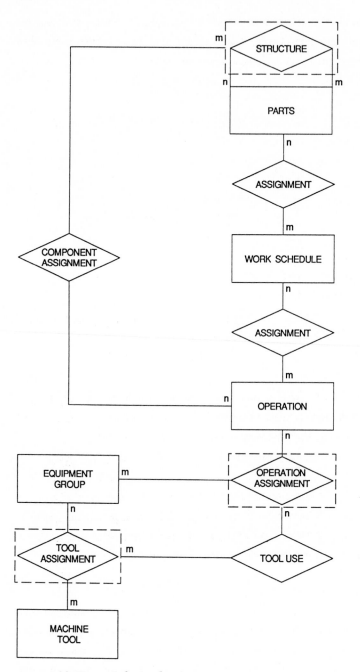

Fig. B.I.03: Entity Relationship Diagram

the structural relationships between bills of materials and operations are specified as data relations. A particular operation can be carried out using several different equipment classes, and in general a particular equipment class can be used to implement several operations. Similar types of equipment are classified together in equipment groups.

Operation assignment is the essential source of information within the production data, on the basis of which production times, refitting times, etc. can be allocated on a machine basis.

The tools to be used can be assigned to the technically relevant equipment groups, and, conversely, technically possible tools can be defined for a given equipment group.

At an early stage the complexity of this data structure demanded the use of comfortable **data management systems**. For this reason bill of materials processors which could use address chaining to generate non-redundant n:m relationships became the starting point of the general **database systems** which are now widespread.

To summarize, primary data management of bills of materials, work schedules and equipment groups is not only the basis of every computerized production planning and control system, it is also the basis for product pricing. Here costs are estimated for individual parts and materials, production times are evaluated from a breakdown of work schedules and, using hourly machine rates for particular equipment groups, these costs too are passed on to the next production level up to the final product level.

2. Planning Levels

The levels of a PPC system have already been represented in the left fork of the Y-diagram of Fig. A.01. The order handling system constitutes the link between the production and sales areas. Such a system does not always belong within production planning and control. In an inventory based serial manufacturing system, for example, the problem of customer order handling can be separated from production order based planning and control. The more customer-specific demands affect the manufacturing process, however, the closer the link between these two systems must be. In the context of **order handling** customer orders are accepted, delivery dates arranged, reservations made and the necessary input data for creation of the production program ascertained. In the case of customer-demand-oriented production (individual orders, variant production) cost information must be established in the course of order acceptance. This preliminary pricing necessitates access to primary production data (bills of materials, work schedules, equipment data).

Master production planning establishes the requirements for the output levels to be produced in the next period in terms of end products, end product groups and independently marketed replacement parts. Here, actual customer orders in hand provided by the order handling department and estimates of expected sales figures are required. As Fig. B.I.04 shows, this planning level determines the sub-areas of material

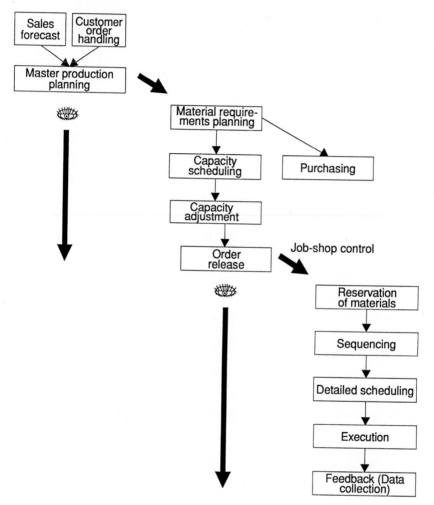

Fig. B.I.04: Multi-stage modelling for order processing

and capacity management. To put it another way, errors that arise in the specification of the master plan, also affect the planning quality at subsequent levels. It is typical of present PPC systems that too little attention is paid to the specification of the master production plan. Hence many systems lack the necessary forecasting support; simulation models for early detection of material and capacity bottlenecks and optimization models for establishing the cost and revenue effects of alternative production programs. One reason for this inadequate planning support may be that in many enterprises the preparation of the master production plan is left to the sales department, which then establishes forecast values in areas where it has inadequate knowledge. The consequence is that considerable disruption is caused in the area of production planning and control, where rush orders and forecast changes require frequent re-planning.

Material management first breaks down the master production schedule data relating to end products into assembly groups, individual parts and materials. Here the data structure is determined from the bill of materials, which describes the composition of the end products from its components. In a deeply layered production hierarchy the breakdown of the bill of materials is a central area of application for computer systems in production. Ranking according to planning levels ensures that each part, although required by various higher level components, need be processed only once.

Incorporation of inventory levels allows **gross/net calculations** to be made. Using lot-sizing formulae production orders for in-house produced parts as well as the need for purchased parts can be established. The need for purchased parts can then be passed on to the purchasing department, and production orders can be given to capacity management.

The initial task of **capacity management** is to carry out capacity scheduling, in which production orders and work schedules are combined. Individual operations are hereby assigned to specific equipment groups, and, on the basis of the master scheduling carried out by material management, operations can be timetabled and production capacity assigned. The result of this step is the so-called capacity overview diagram, in which, for example, the load on a particular equipment group can be represented in bar chart form (see Fig. B.I.05).

If capacity bottlenecks occur (as in period 3 of Fig. B.I.05) various adjustment procedures can be undertaken in the course of **capacity adjustment**. These include:
- introduction of overtime or extra shifts,
- relocation of operations using critical equipment groups to functionally equivalent equipment groups with spare capacity,
- increased production intensity, and
- temporal relocation of operations to periods where the load on the equipment in question is less.

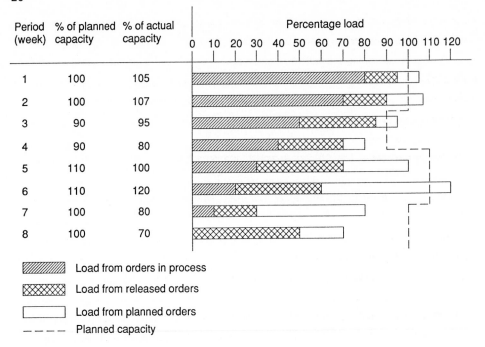

Period (week)	% of planned capacity	% of actual capacity
1	100	105
2	100	107
3	90	95
4	90	80
5	110	100
6	110	120
7	100	80
8	100	70

Load from orders in process
Load from released orders
Load from planned orders
– – – – Planned capacity

Fig. B.I.05: Capacity load overview for one equipment class

Although the first measures have no effect on the order flow, this is unavoidable in the case of temporal relocation. It also implies a temporal effect on preceding and subsequent processes. In the case of complex interconnections this temporal interdependence can lead to costly re-planning problems. For this reason, computer systems that aim to resolve this problem by the use of simple **priority numbers** or heuristic algorithms have often failed. In addition, the stability of planning outcomes is relatively low when there is such susceptibility to disruption, because with precise temporal planning of operations replanning will frequently be required. This has led to the present situation, in which solutions based on the organizational capacities of humans in the framework of a man-machine interaction are favored over batch or algorithm based solutions. It should, of course, be noted that the basic planning complexity can also overstretch human organizational abilities, so that a promising development might be the combination of both approaches, that is, the adoption of planning algorithms within an organizational dialog, for example through the use of expert systems.

Order release transfers orders from the planning to the implementation phase. This merely involves an examination of a section of the planning horizon. Before an order is released an availability check is carried out with respect to the necessary components, machines, tools and labor. For automated production the availability of NC programs must also be ascertained. Once availability is determined those orders whose planned starting date falls within a certain specified period are forwarded to production.

Within **production control** released operations are assigned to equipment groups in accordance with new optimization criteria. Such criteria might be:
- avoidance of waste by optimized cutting,
- avoidance of refitting costs,
- production technology requirements, such as even load on particular machines.

Here too, new computer architectures could be demanded to cope with the variety of requirements as seen by the enterprise. In the planning context, computer systems are usually employed on the basis of business criteria. At the same time, the operation of the computer system tends to be based on "office" hours, i.e. interactive processing is carried out in a one or two shift operation. In the context of production control, however, there is a close relationship with production itself, which often works on a two, three, four or even five shift basis. At the same time it requires extreme flexibility of the computer system in its link-up with various peripherals. These arguments have led production control to change over to more process control oriented hardware.

In the course of production control, actual performance data are recorded as part of the **operational data collection**. Detailed entries are made concerning:
- order-related data (production times, output levels, qualities),
- machine-related data (breakdown, run times, interruptions, maintenance measures),
- personnel-related data (attendance times, hiring and firing), and
- material-related data (stock movements).

The data included in the operational data collection are not only prerequisites for up-to-date production control, they also constitute the infrastructure for various areas of application. Personnel related data are also needed for gross wage calculation; current order related data are needed for continuous price estimating. For these reasons, planned/actual deviation analysis can continuously assess both output and costs, so as to allow speedy corrective measures to be taken in the course of the production process.

Data related to completed products are transferred to **dispatch control**, which can then optimize packaging, itineraries, etc.

The operation of a production planning and control system therefore pursues the entire **logistic chain** from order acceptance through material requirement planning, purchasing and manufacturing to dispatch. At the same time, the necessary data are produced for the business functions of financial accounting, cost accounting and salary calculation. Fig. B.I.06 depicts the structure of such a chain in a typical functionally oriented business organization and makes it clear that the logistic chain runs at right angles to such a functional organization. Currently PPC systems in a typical industrial enterprise account for about 60% of the total information processing transactions. This indicates the extent of the influence of PPC systems on the operational structure.

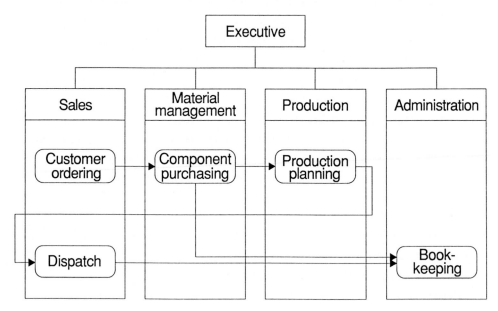

Fig. B.I.06: Logistic chain

3. Degree of Implementation

Fig. B.I.07 depicts the extent of implementation of the various planning stages via the width of the relevant boxes. The width of any stage indicates the extent of the practical implementation in terms of in-house developments or standard software. It is clear from Fig. B.I.07 that PPC systems have significant order processing content. Master planning, which establishes the production program in light of capacity and material supply constraints, receives in contrast little support, and the implementation of the planning stages diminishes as their practical applicability increases.

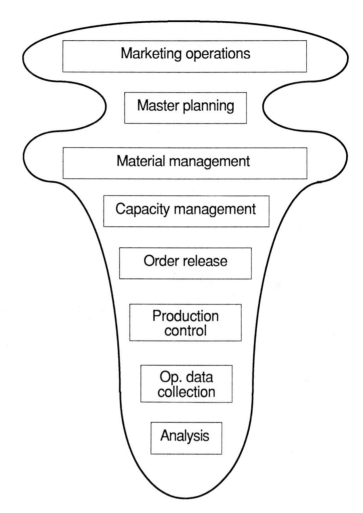

Fig. B.I.07: Current weighting of PPC planning stages

So it is clear that up to now **generality of planning and control**, which is the obvious aim of any PPC system, has scarcely been attainable either using standard software or in-house developed systems.

In addition to the lack of generality, the concept of **successive or serial planning** causes problems, because the results of one stage are the input of the following stage and feedback can be achieved only with difficulty. In principle, though, it is easy to think of examples in which this feedback is essential, e.g. between material and capacity management. In addition to these conceptual failings, the system also has shortcomings in its restriction to specific production structures.

Figs. B.I.08,a and B.I.08,b show two extreme **manufacturing types**. Fig. B.I.08,a depicts a **raw materials based manufacturing structure**, in which a multiplicity of final products are manufactured using a small set of raw materials. This is typical of the ceramic, paper, chemical and food industries, where final products are often only distinguishable through packaging or size differences. In this type of production primary considerations are the progress of material flows, the balancing of assembly lines and optimization of the production sequence, given refitting costs. In general, processing intensity is not very pronounced: individual jobs may be integrated as closed assembly lines - in extreme cases there may only be a single production facility.

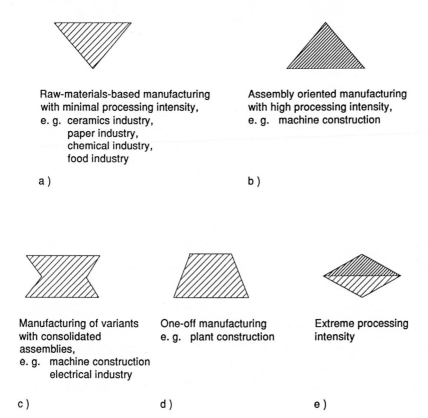

Raw-materials-based manufacturing
with minimal processing intensity,
e. g. ceramics industry,
         paper industry,
         chemical industry,
         food industry

a)

Assembly oriented manufacturing
with high processing intensity,
e. g. machine construction

b)

Manufacturing of variants
with consolidated
assemblies,
e. g. machine construction
         electrical industry

c)

One-off manufacturing
e. g. plant construction

d)

Extreme processing
intensity

e)

Degree of shading indicates the suitability of existing PPC systems

Fig. B.I.08: Manufacturing types

In contrast, Fig. B.I.08,b shows the dominating manufacturing structure typical of manufacturing industry, e.g. mechanical engineering. A variety of materials and purchased parts are combined using a variety of manufacturing and assembly procedures to create complex final products. Here, the management of product assembly and manufacturing instructions in the form of bills of materials and work schedules, and consequently the determination of production orders via demand based procedures are the overriding considerations. From the basic types B.I.08,a and B.I.08,b further forms can be derived, as shown in B.I.08,c to B.I.08,e. Many German industrial enterprises are typified by strong customer orientation requiring a great variety of products, often very specific, one-off designs. Fig. B.I.08,c depicts such a case: here, at first, parts are collectively assembled, then specialized finishing of the end product is introduced to handle variant problems.

In Fig. B.I.08,d the one-off production character is even more pronounced, in that there is no well-developed assembly grouping. Fig. B.I.08,e depicts an extreme processing intensity, in which raw materials are used to manufacture various intermediate products, which are ultimately assembled to create complex units.

The PPC philosophy presented above primarily supports the manufacturing form B.I.08,b, while the other structures are inadequately handled. In particular, there has long been a need for special systems for one-off producers, in which the relationship between a production order and a customer order can be maintained at all production levels. A general assessment of the appropriateness of the PPC philosophy is indicated in Fig. B.I.08 by the intensity of the shading. As a result of these factors, few industrial enterprises are fully satisfied with their planning and control systems. A significant implementation problem arises, for example, where PPC systems are installed which are inappropriate to the manufacturing structure of the enterprise. But there are also many enterprises which have become stuck in their implementation of a basically suitable system because the organizational costs have been underestimated.

Despite these failings, the extent of the system integration achieved by the use of database systems should not be misunderstood. These systems are in a position to process large volumes of data for bills of materials, work schedules and orders. Also the reduction in the complexity of problems through the use of a layered planning concept should not be underestimated.

4. New Approaches to PPC Systems

Several of these limitations of existing PPC systems were recognized at an early stage. Hence, for several years PPC systems have been available for customer-oriented order production, which provide particular support for the process of order specification, in which, for example, order-related bills of materials are derived from component master

32

files, and production and purchasing orders can be defined without the need for a complete bill of materials specification. These systems, however, adhere closely to the existing planning architecture and will, therefore, not concern us further. On the other hand, extensions can be detected which express an initial careful modification of this architecture, which, however, falls short of criticism. This concerns the principle of layered planning, in that it demands simultaneous material and capacity management on the one hand, and, on the other, a stronger link between production planning and control in the context of load-oriented order release.

4.1 Simultaneous Material and Capacity Management

The links between material and capacity management are inadequately handled in successive planning models. For this reason new extensions have been created within an

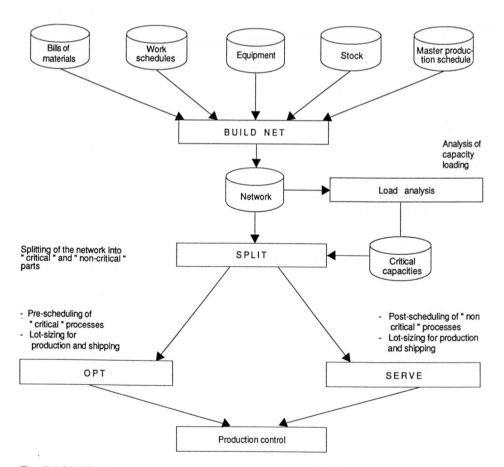

Fig. B.I.09: The OPT/Serve system

otherwise traditionally oriented PPC architecture to take account of this state of affairs. The system OPT (see Fig. B.I.09), currently being marketed very aggressively, breaks down the total order network into those production orders which could put pressure on equipment tending to bottlenecks, and those orders requiring organizationally unproblematic capacity units (see *Smith, OPT-Realisierung 1985*).

This splitting up according to order type allows reductions in network size and hence in planning complexity. Critical orders are handled first using a form of **forward scheduling** and are thereby assigned higher priority relative to other orders. After this initial planning **backward scheduling** fits the non-critical orders into the critical order schedule. Although the algorithm of the OPT system is only imperfectly disclosed, the basic idea seems thoroughly sensible. For this reason it has already been adopted by other PPC systems.

Several new PPC systems offer quasi-simultaneous capacity and material management at the level of strongly **event-oriented scheduling**. This means that concrete order scheduling is carried out by breaking down the specific orders into their required components, taking material requirement functions (i.e. delivery dates for the required components) into account in the capacity control scheduling (see *Kazmeier, Belastungssituation im Rahmen eines PPS-Systems 1984*). This procedure is of course only possible within an event-oriented approach. In addition, there is inadequate algorithm support for the assignment of priorities to critical materials and capacities.

4.2 Load-Oriented Order Release

In traditional functionally structured PPC systems orders are released from the planning to the scheduling level, in that orders lined up for a particular planning period are transferred to production in accordance with their **planned** starting date, once availability checks have been made. The release criterion, therefore, is the date specified by **planning**, which is forwarded to control in accordance with this stepwise approach. Since this takes inadequate account of the capacity situation (availability checks are usually restricted to materials and components) it can result in production overload, and consequently excessive stocks in progress and processing times. Here the concept of load oriented order release is relevant (see Fig. B.I.10). According to the "filter principle" only those orders are released which can be processed given the capacity situation (see *Wiendahl, Verfahren der Fertigungssteuerung 1984*). This necessarily involves a relaxation of the batch principle, since order release now involves "longer-range" consideration of the requirements of subsequent batches (see Fig. B.I.04), and can no longer be guided exclusively by stratified criteria.

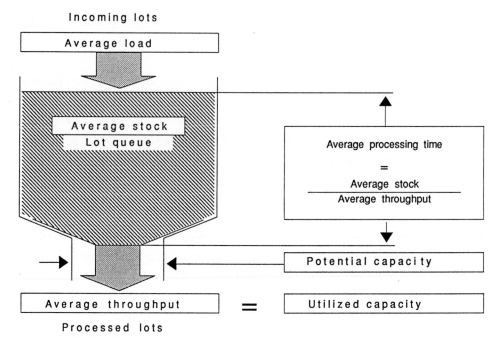

Fig. B.I.10: Funnel diagram
    Source: *Wiendahl*

4.3 Kanban

In recent years the Kanban principle, which originated in Japan, has led to sometimes exaggerated expectations, in which a supposedly far-reaching simplification of the production organization, in comparison with complicated EDP systems, seemed attainable (see *Wildemann, Flexible Werkstattsteuerung 1984*). The Kanban principle prescribes a "minimum inventory level"-oriented production approach, in which a preceding production level generates new production orders when it observes that its inventory level has fallen below the minimum level. A simplified organization is introduced, in which previously specified output levels are produced and which are oriented to the transport containers (Kanban containers). Each Kanban container is also assigned an order card (Kanban means card in English), on which the order quantity and other details are noted. The planning process is executed by transfer of this card (similar to the well-known commuter card procedure). The Kanban principle is in general organised as a pick-up system: order prescriptions from the prior production level determine the further absorption of output quantities into the production process.

In contrast to the MRP principle (Material Requirement Planning), Kanban is a minimum inventory level system, while the requirement controlled organization of MRP is strictly a zero-inventory system, since orders are only produced when a specified need exists, and not when a minimum level has been reached. In spite of this Kanban is often discussed as an inventory reducing system since it leads to an acceleration of the production flow. Furthermore, the discussion of Kanban has led to greater consideration of the question of refitting, and in certain cases has caused considerable organizational improvements. In addition to internal applications of the Kanban principles this procedure can also be applied to transportation between firms. Particularly impressive is the Japanese example involving automobile manufacturers and supply firms, where supply firms can organize their deliveries on a strict hourly schedule.

Implementations have shown however that, for various reasons, Kanban cannot be employed as a general control system in German industry. In the case of inter-company applications, it presupposes a very close involvement between producers and suppliers, which in Germany is not generally the case; and within the enterprise it assumes stability of quantities produced and high quality reliability.

Nevertheless, the control principle can be successfully applied to production sub-areas in German industry.

4.4 Running Total Concept

For assembly-oriented serial production, typical of the automobile industry, the running total concept seems to represent a new planning system development. A running total refers to a cumulative value which can relate to various standard units. If the running total is based on planned units it is a planned running total. Correspondingly realised values are referred to as actual running totals. Fig. B.I.11 shows a typical running total as planned and actual values. The planned value represents the cumulative planned production quantities for a particular part, the actual running total corresponds to the actual manufactured output. The figure also makes a further aspect of the running total concept clear: the cumulative values always relate to a specific point in time.

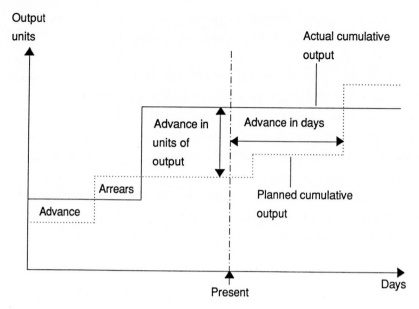

Fig. B.I.11: Running total diagram

Comparison of planned and actual running totals can yield further insights. If the solid line lies above the broken line production is in advance of schedule. This advance can be expressed either in quantity units through the vertical distance, or in time units through the horizontal distance. In Fig. B.I.11 this is expressed relative to the present by the arrow signs.

A logistic concept based on running totals can access a multiplicity of reference groups for which running totals can be calculated.

For example, these may be:

Planned	Actual
Call-off running totals	Dispatch running total
Production plan running total	Assembly running total

Inventory level figures are given by the difference in the running totals of inventory additions and inventory withdrawals.

The concept presumes a close customer relationship, and so the starting point is the cumulative record of customer orders, where agreed delivery dates constitute the basis of time allocation. This **orders running total**, which is a planned value, can be compared with the **dispatch running total**, which is the corresponding actual value.

The running total concept is widespread in the automobile industry. Here planned order totals are compared with effective call-up as other running totals. The system can embrace the entire logistic chain down to a process-specific production overview. The differences between actual and planned values, and their interpretation as advance and arrears (expressed either in quantity or time units), make clear-sighted control of the enterprise possible. In particular, the consequences of changes can easily be highlighted. The running total concept, therefore, represents an effective extension of present production planning and control methods. In determination of planned running totals, for component requirements, for example, it makes use of standard requirement assessment methods. The method gives rise to a need for evaluation and information support in the area of production planning and control. It has additional relevance in the area of inter-company data exchange. In the automobile industry manufacturers and suppliers already exchange running total figures for deliveries, orders and call-offs.

4.5 MRP II

Developed by Oliver Wight, the concept MRP II (meaning management resources planning, as compared to MRP - material requirement planning) locates the planning and control problem within the totality of a logistic chain. Here, weight is given to hierarchical

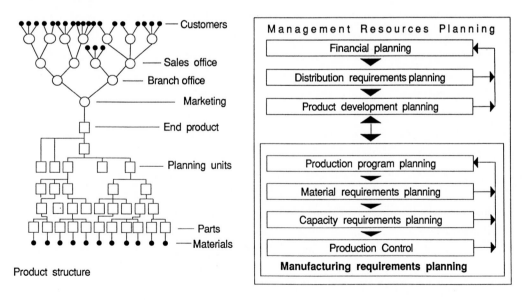

Fig. B.I.12: MRP II model
Source: *Gesellschaft für Fertigungssteuerung und Materialwirtschaft e.V.*

planning notions, from strategic planning through production of master plans down to production plans (see Fig. B.I.12).

4.6 Decentralization

On the basis of the intensified development of control units arising from the new information technology, PPC systems must increasingly concern themselves with questions of organization of control centers, control of flexible production systems, etc. This gives rise to small independent feedback loops, which must, however, be tied to higher level supply systems. For the moment this statement of the problem must suffice: the entire question will be considered much more intensively in the subsequent discussion of CIM process chains (see Section C.III.a).

b. Computer Aided Design (CAD)

1. Tasks

The task of the design department can be subdivided, according to guideline 2210 of the VDI (Verein Deutscher Ingenieure - the association of German engineers) into new design, adapted design, variant design and fixed principle design. The design process can be broken down into the following phases:

1. conception (specification analysis, compilation of solution variants, assessment of the solutions),
2. development (specification of the solution concept, scale design, model construction, assessment of the solutions),
3. detail (representation of individual parts, assessment of the solutions).

The three levels are, of course, interdependent and, depending on assessment results, can be repeated cyclically. The third level forms, with the preparation of production instructions, the transition to process planning. The important activities in the design context are:

- obtaining information from files and existing data stocks,
- calculation of guaranteed loads and tolerances to be observed,
- preparation of drawings,
- undertaking of technical and economic analysis of the design.

The typical working method of the designer is functionally-oriented, i.e. starting from the statement of the task, functional elements are chosen and put together to create a solution. In the conception phase, in which a functional plan and basic sketch are produced, about 71% of the documents will be in graphical form (see *Spur, Krause, CAD-Technik 1984, p. 257*), during the development phase it is 95%, and in the finalising phase it is 65%. Hence the significance of computer-supported graphic production is already obvious. With the recognition that progressive automation of the design and development phase are increasingly decisive both for production and managerial processes, and hence for the cost of their product, their functional scope increases. This is assimilated within the CIM concept. In what follows, however, CAD will be regarded primarily as a data processing function, closely related to the graphically-oriented design functions. The extent of computer support is greatest in the development and detailing phases, and still has a relatively minor role in the conception of a new product. Extensive computational methods (finite element method, load simulation, etc.) are referred to under the term Computer Aided Engineering.

2. Geometric Models

Three different computer-internal representation of geometrical objects can be distinguished (see *Hübel, Datenbankbasierter 3-D-Bauteilmodellierer 1985*):
- edge representation,
- surface representation,
- volume representation.

Edge-oriented models (see Fig. B.I.13,a) represent objects using points and contours. This representation is primarily suited to two-dimensional geometries. Although edge models can be used to generate spatial representations, these tend to be unclear. It is also impossible to show sections or shadowing. **Surface-oriented models** (see Fig. B.I.13,b) geometrically depict the "skin" of a body using surfaces. The surfaces have boundaries which are defined by two neighboring, touching or intersecting surfaces. Points are the result of the intersection of three surfaces, or constitute definitions for establishing the location of contour elements. A **volume-oriented model** is constructed from set-theoretic assembly of various volume-oriented basic forms. In the example shown in Fig. B.I.13,c amalgamation and difference operations are represented. For all points, the position relative to the object can be determined, regardless of whether they lie within or without the object. Projections can generate a variety of views using such a computer-internal spatial model. Since they are all generated from the same model, the views are logically consistent and integrated - unlike the separately generated views of a 2-D model. Given the restricted possibilities of the surface-oriented models, volume-oriented representations tend to be employed.

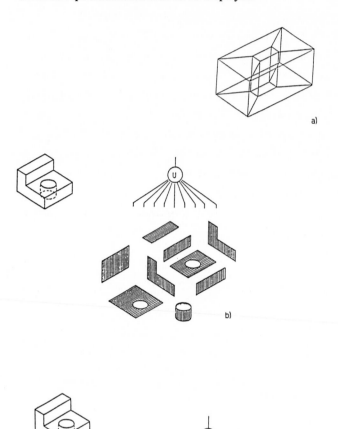

Fig. B.I.13: Models of CAD representation schemes
from: *Hübel, Datenbankorientierter 3-D-Bauteilmodellierer 1985*

These models also constitute the starting point for calculation and simulation experiments. If only a two-dimensional representation is required (e.g., for representing circuit diagrams) then line or surface models are justified. Alongside 2-D and 3-D systems, there also exists the so-called 2 1/2-D system. Here a 2-D view is described and stored, and using mathematical operations (rotation and translation) a 3-D model can be created.

3. CAD Standard Interfaces

Since the computer internal representation of geometric elements differs between the various CAD systems (even for the same geometric model) the direct exchange of data between CAD systems is problematic, as there is not necessarily a 1:1 relationship between the elements of the various data models, but rather 1:n or even n:m relationships are possible. For this reason standard interfaces have been developed which first transform the data of a CAD model into "standard format", such that they can then be converted to the system (see *Sorgatz, Hochfeld, Austausch produktdefinierter Daten 1985*). Well-known interfaces are the IGES format (IGES = Initial Graphics Exchange Standard) which was issued as an ANSI norm as early as 1981, the French system SET and VDA-FS, the free-form surface interface developed by the *Verband der deutschen Automobilindustrie* (*VDA* - the German automobile industry association) (see *Grabowski, Glatz, Schnittstellen 1986*). The conversion of the drawings of a CAD system into standard format is carried out by a pre-processor belonging to the sending system, and a post-processor transforms standard format into the drawing format of the receiving CAD system (see Fig. B.I.14).

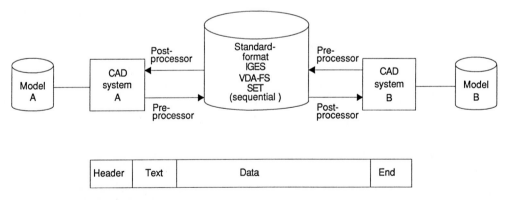

Fig. B.I.14: CAD interfaces

For two CAD systems to be able to communicate with each other, therefore, each must be equipped with the necessary pre- and post-processors for handling standard format in both directions. The interface itself is a relatively simple construct. The IGES system, for example, consists of sets of 80 characters each in ASCII format. Each set is of the form: Start, Global Directory Entry, Parameter Data, Terminate. IGES was initially developed for 2-dimensional data models, and later extended to handle 3-dimensional line models. Exchange of volume data is not possible.

The fundamental nature of the problem is represented in Fig. B.I.15. It shows the relationships between VDA-FS interface data models and the CAD system STRIM 100 (see *Rausch, de Marne, VDA Flächenschnittstelle 1985*). A multitude of logical relationships exists between the elements of the data models. For example, the element "master dimension" is represented as an isolated point in the VDA-FS model and as an isolated line in the STRIM 100 model. Fig. B.I.16 illustrates this more clearly with a simple example. The arrow which exists in one CAD system does not exist in the second CAD system and hence must be reduced to its component parts.

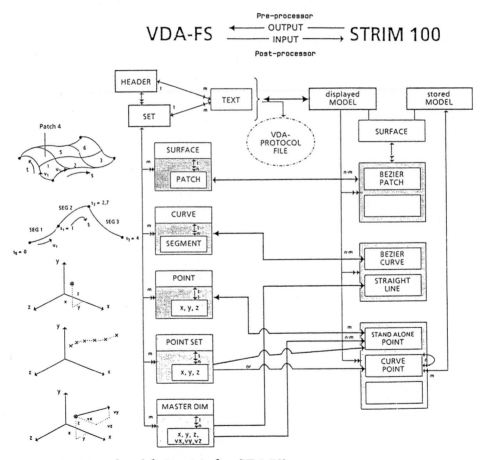

Fig. B.I.15: VDA surface definition interface (VDA-FS)
from: *Rausch, de Marne, VDA-Flächenschnittstelle 1985*

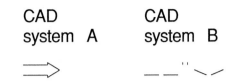

Fig. B.I.16: Representation schemes in CAD systems

The non-conformity between data models is, therefore, one cause of the considerable loss of information that can occur in the course of transformation via standard interfaces. For this reason the transferred data sometimes need to be manually reworked.

The VDA free-form surface interface developed by the *Verband der deutschen Automobilindustrie* is used primarily for the exchange of surface data, which is suited to the production of tools for body-making (see *Encarnação, CAD Handbuch 1984, p. 53*). Although interfaces still display considerable failings, they nevertheless point the way to data exchange between different CAD systems within the enterprise, and in inter-company communications, e.g. between manufacturers and suppliers in the automobile industry (see *Schwindt, CAD-Austausch 1986*).

4. Interactive Control

Computer supported design was planned from the outset as a highly interactive process. At all design stages the modification of solutions through an interactive decision process stands in the foreground. Economic factors are included and technical alternatives assessed in order to attain the "optimal" solution. Extended processing times for the completion of mathematical operations or the construction of complicated 3-dimensional figures may require a batch processing approach. These would be started as interactive processes, then run as background jobs on the CAD machine.

Appropriate instruments are installed at the user interface between the designer and the EDP system, such as graphic tablets, light pens and mouse techniques. Fig. B.I.17 shows the use of a graphic tablet, whose functions can be activated via a light pen. The graphic tablet shows specifications of standard components (contours, etc.), which can be copied onto a current drawing. A high resolution display (approx. 600*800 to 1000*1000 pixels) must be available for representing drawings.

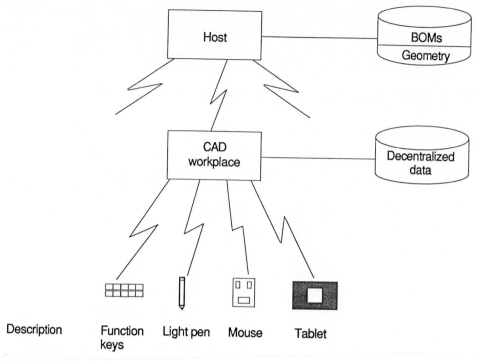

Fig. B.I.17: Structure of a CAD workplace

The CAD area is particularly well suited to the introduction of workstations. For purely technical reasons the use of graphics terminals caused an early prevalence of workstations, where these were employed either as stand-alone systems or linked to central computers. It must be recognized that the workstation will gradually assume more and more processing functions. On the one hand this will affect the drawing preparation function, on the other it will support the design process itself. The tendency is for the central computer simply to house the management of geometric and bill of materials information, while almost all CAD functions are transferred to the workstation - down to extremely extensive computing functions. The engineering workstation is, in general, a powerful microcomputer.

Apart from the utilization of the high resolution graphics terminals they offer, the essential reasons for the location of the processing functions in workstations are to relieve the load on the central computer, to improve user accessibility and to support decentralized organizational measures. With respect to their hardware properties, the ability to link up various plotting and input media (light pen, mouse, graphics menus, tablets, etc.) is crucial.

Alongside the tendency towards relocation of functions from large-scale CAD systems to workstations, there has been an increased availability of so-called low-cost CAD systems, developed from the outset for microcomputers (e.g. AutoCAD, CADdy, VersaCAD) albeit only over standard interfaces (file transfer, terminal emulation) which can be linked to a central computer.

c. Computer Aided Planning (CAP)

The work schedule describes the transformation of production parts from their raw to their finished state. The starting point may be a single material, or even, in the case of assembly-based operations, in-house produced assembly parts and components. The work schedule contains the sequence of operations for the production of a part, allocates the equipment for the operations, specifies standard times and wage groups. The bases of the work schedule are the geometric and technical specifications. Geometric specifications are taken from the drawings produced by the design department. They may also contain technological data, e.g. concerning material properties, tolerances, surface properties of the production part, etc. Sometimes, however, the preparation of technological information is a subject of the work scheduling process itself. Bills of materials are also important work scheduling documents. Computer supported work scheduling needs to distinguish whether the work schedule is being created for a traditional manufacturing process, or for a computer controlled production unit (NC machines). In the latter case the work schedule will be replaced or supplemented by NC programs.

1. Work Scheduling for Conventional Processing

For conventional manufacturing processes a work schedule such as that already shown in Fig. B.I.01 is created. Alongside design data, drawing and bill of materials, access is made to various other data sources in the production area (see Fig. B.I.18).

By searching for similar parts in a parts database, work schedules can be created by accessing other work schedules containing essentially the same basic operations which merely need to be modified. Standard work schedules, which are typical of certain parts groupings, can also be used as a starting point for the creation of a specific work schedule. Here the management of bills of materials and work schedules in itself constitutes an important information function for the creation of new work schedules.

46

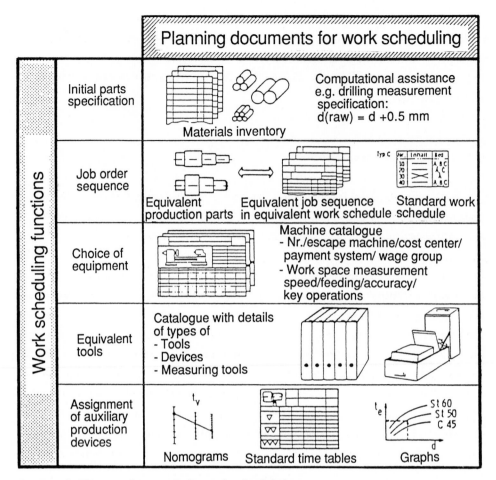

Fig. B.I.18: Planning documents for work scheduling
from: *Spur, Krause, CAD-Technik 1984, p. 445*

Other primary data of a PPC system may also be used (see Fig. B.I.03). In choosing the materials to be used, access can be made to the materials catalog, in which qualities such as strength, surface characteristic, castability, weldability, and stretching limits are contained. Which machine to use is established with reference to data from the equipment group, and the choice of machine tools on the basis of machine tool data. Determination of standard times is based on nomograms, standard times tables and diagrams. This kind of information can, of course, be stored electronically, but typical storage forms are microfilm and paper documents.

Work scheduling is carried out by process planning. This is centrally organized, at least at the factory level, but sometimes also at the level of the enterprise responsible for several factories. To make greater allowance for the close relationship between design and production, attempts are increasingly being made to forge closer organizational links between process planners and designers. This can be seen as an attempt to break down existing functional divisions between design and process planning. This tendency is even stronger in the case of work scheduling for computer controlled production units.

2. Work Scheduling for NC Machines

Computer controlled production units are machine tools or hand operated machines, for which path and switching information is translated by computers into the appropriate movement and switching instructions. This path control implies that the calculation of the tool path along a specified part contour (geometry data) is taken over by the computer. Further examples which extend the degree of computer application, in addition to the classical NC (Numerical Control) machines, are CNC (computerized NC) machines, DNC (Direct NC) machines and robot systems. The definition of these systems will be extended below.

Whereas in a conventional production unit the machines are controlled by humans on the basis of information contained in the production documents (work schedules and drawings), in a computer controlled production unit these functions are performed by computer programs. Hence the programs take over the functions of both the information sources "work schedule" and "drawing" and of the machine controller who translates these informations into concrete control measures. An NC program can, therefore, be seen as a detailed work schedule which goes beyond the information function, however, to incorporate instructional functions.

The database required for NC programming is similar to that used in manual production planning. The basic information consists of geometry data, such as specification of the cutting path of a tool, technological data, such as chuck or intersection values, tool data, equipment data and material data. The logical structure of NC program creation is shown in Fig. B.I.19, which also indicates the uses of, and extension to, the data structure.

1	PARTNO/D-AXLE	**General declarations**
2	MACHIN/PP1	
3	MACHIN/ZEISIG	
4	MACHDT/30. 120 0.1. 3.5. 3000. 0.8. 200	

5	CONTUR/BLANCO	**Raw material description**
6	BEGIN 0.0 YLARGE. PLAN. 0	
7	RGT/DIA. 100	
8	RGT/PLAN. 330	
9	RGT/DIA. 0	
10	TERMCO	

11	SURFIN/FIN	**Parts description**
12	CONTUR/PARTCO	
13	L 1 = LINE/50. 25. 90. 30	
14	FASE = LINE/ (POINT/310. 22) .ATANGL. 45	
15	MO. M1. BEGIN/20.0.YLARGE.PLAN.20.BEVEL.3.ROUGH	
16	RGT/DIA.40.ROUND.2	
17	LFT/PLAN.50	
18	RGT/L1	
19	FWD/DIA.60.ROUND.4	
20	LFT/PLAN.130	
21	RGT/DIA.70.ROUND.2	
22	LFT/PLAN.160	
23	RGT(LINE/160.40.210.45)	
24	M2.FWD/DIA.90	
25	RGT/PLAN.230.ROUND.2	
26	LFT/DIA.80	
27	RGT/PLAN.250.ROUND.4	
28	LFT/(LINE/250.35.280.30)	
29	RGT/PLAN.280.ROUND.2	
30	LFT/DIA.50	
31	RGT/FASE.ROUGH	
32	M3.RGT/PLAN.310.ROUGH	
33	M4.RGT/DIA.0	
34	TERMCO	

35	PART/MATERL.203	**Technological definitions**
36	CSRAT/60	
37	CLDIST/2	
38	OVSIZE/FIN.1	
39	PLANE = TURN/SO.CROSS.TOOL.100.1.SETANG.180.ROUGH	
40	SCHRU = CONT/SO.TOOL.200.2.SETANG.110.ROUGH.LONG	
41	SCHL1 = CONT/SO.TOOL.200.2.SETANG.110.FIN.OSETNO.7	

42	CHUCK/11.50.200.30.104.55	**Control instructions**
43	CLAMP/50	
44	COOLNT/ON	
45	WORK/PLANE	
46	CUTLOC/BEHIND	
47	CUT/M3.TO.M4	
48	WORK/SCHRU.SCHL1	
49	FDSTOP/PLANE.207	
50	CUT M3.RE.M2	
51	FDSTOP/NOMORE	
52	CHUCK/222.50.200.30.94.70	
53	CLAMP/245.INVERS	
54	WORK/PLANE	
55	CUTLOC/BEHIND	
56	CUT/M1.RE.MO	
57	WORK/SCHRU.SCHL1	
58	CUT/M1.TO.M2	
59	WORK/NOMORE	
60	FIN	

Fig. B.I.19: Structure of an NC program

d. Computer Aided Manufacturing (CAM)

The term CAM is not used consistently. Sometimes it refers simply to the control of computerized conveyance, storage and production machines; sometimes it is defined very broadly to include production control functions, such as are included here in the production planning and control concept. The use of the term in the following discussion adheres more closely to the first variant. The concept of Computer Aided Planning (CAP) is thereby closely related to Computer Aided Design (CAD) as well as to the CAM applications which are nearest to control.

1. Automated Production

1.1 Machine Tools

NC machines were the starting point for computerized production (see Fig. B.I.20,a). Programs were entered into the production machines, (e.g. lathes, milling and drilling machines) using paper tapes (see *Kief, NC Handbuch 1984*). The control itself was via fixed wiring. This means that even control adjustments were scarcely possible. In addition, changes in the NC programs could only be effected by replacing the paper tapes. CNC machines (CNC = Computerized Numerical Control) were developed to allow greater operating flexibility. With CNC machines (see Fig. B.I.20,b) a small computer, usually a microprocessor, is attached to the machine tool and takes over the tasks of numerical control. Because of this programs can be entered directly into the machine, and hence changes can be implemented much more easily. The programs may also be entered on paper tape, as with NC machines, but can then be stored and processed.

In a DNC (Direct Numerical Control) system (see Fig. B.I.20,c) several NC or CNC machines are linked to one computer, which manages the control programs (the NC program library) and sends them at the appropriate times to the individual machines. Programming and amending of NC programs can also be carried out in this central computer. In addition to the control of production units, a DNC computer can be employed for evaluation and data collection tasks (e.g. machine statistics).

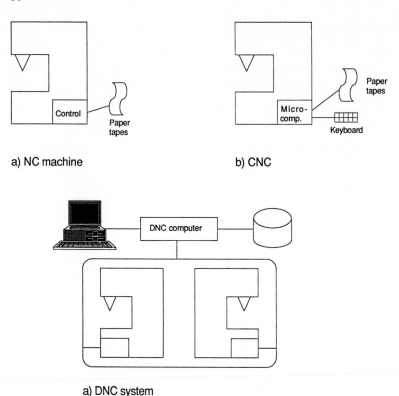

a) NC machine   b) CNC

a) DNC system

Fig. B.I.20: Control of NC machines

1.2 Robots

The automated control of hand operated tools and robots presents similar problems to those of machine tool control (see *Reitzle, Industrieroboter 1984*). The programming, however, exhibits special features, in that, as well as the use of programming languages, so called play-back and teach-in programming are possible. In **play-back programming** a robot is moved by manual manipulation of its tool holder or carrier. This movement is stored and an applications program converts it into robot control, which can then be called up as often as desired. In **teach-in programming** a movement is carried out using switch and key operations, which can then also be stored for future access.

In contrast to normal NC programming, the integration of sensory messages and sensory data requests constitutes a special feature. Robots are equipped with the sensory capacities to recognize and react to the properties of production parts and tools. Hence with robots logic up to the level of artificial intelligence techniques (expert systems) can be implemented. With respect to the programming of robots, the same statements about the link between geometric and technological information apply as were made above in a more general context. In the PROCIM system, for example, automated robot programs are created from geometry information about the production part to be produced and positioning and cursor data of the parts to be assembled.

1.3 Storage Systems

Automated production systems also demand automation of the supply of tools, production parts and materials. Here automated stock-keeping systems are increasingly being introduced. Their role is to manage individual stock containers or bays, as well as to control stock movements. Optimizations are carried out, so that, for example, stock additions and withdrawals are effected with the smallest possible number of movements of the supply vehicle. These tasks can be incorporated in a **dedicated control system**. A (higher level) production control system simply provides key data in the form of stock quantities and supply dates to the dedicated system. Here orders are administered and transformed into control instructions in accordance with optimization criteria. The use of dedicated systems is sensible because control needs to be effected contemporaneously, for this reason realtime operating systems must be employed and hence special computers (process computers) are required. The dedicated sub-system then sends summarized data concerning stocks of parts, etc. back to the higher level PPC system.

1.4 Conveyance Systems

The automation of conveyance systems is characterized by **driverless transport systems**. They are conducted along induction loops, which specify the conveyance routes. Positional data relating to vehicles, place of origin and destination, as well as quantity data are needed for control. The stations are assigned to equipment groups and stores. Several transportations can be put together to form a timetable, in which each transportation constitutes one line of the timetable. If transportation is undertaken immediately after production the conveyance system can also be used as an automatic source of information for the operational data collection system, in that it communicates the status information "operation completed".

The combination of automated storage and conveyance systems also opens up new possibilities for production control. Since it is well known that a substantial proportion of total order processing time is taken up by conveyance and lay times, a considerable diminution can be achieved by tightening up internal logistics. If it is always known where an order (job) is, and an automated storage and conveyance system ensures that a specific order can be immediately identified, sought, withdrawn or deposited and transported, then flexible control procedures which respond quickly to short-term events become possible.

2. Computerized Organizational Forms for Flexible Production

As a result of the combination of various computerized production systems new organizational forms have developed. These substantiate the claim that the use of computer systems can infiltrate deep into the organizational structure. On the other hand the use of computers bears fruit only after the creative assimilation of their potential for converting to new organizational structures. It is a common feature of all systems that they tend to aim at greater **functional integration**. Individual organizational forms sometimes constitute stages in the development of increasing integration of automated production facilities; sometimes, however, they are also complementary, parallel developments. In the following discussion some typical organizational forms will be presented (see *Hedrich, Flexibilität in der Fertigungstechnik 1983*).

2.1 Processing Centers

A processing center is a machine that is equipped with NC control and automated tool exchange, and that can handle the execution of several job operations in one run (i.e. in one uninterrupted process). The drilling and milling process can be regarded as a classic example of a processing center. Processing centers are employed in small to medium scale serial manufacturing, and for highly complex production parts they prove economic even at small output levels. Integration of several operations allows the processing time to be reduced considerably below that of the functionally specialized organization.

2.2 Flexible Production Cells

A flexible production cell consists of automated machines, a buffer storage system for production parts and an automated clamping and loading station. Additional computerized functions can provide tool breakage control, tool locking measurement, variable spatial coding for tools and automated standing time monitoring. Hence a flexible production cell is a unified collection of several numerically controlled machine tools, which can automatically process similar production parts over an extended period (see *Hedrich, Flexibilität in der Fertigungstechnik 1983, p. 112*). If the acquisition and deposition of production parts is also automated, then flexible production cells can be regarded as autonomous.

2.3 Flexible Production Systems

A further development of the flexible production cell is the flexible production system. It consists of the processing system, the material flow system and the information flow system, which are all linked together. Total control is executed by a computer which takes over the conveyance of production parts and tools, as well as providing the production facilities with the relevant control programs (NC programs). Fig. B.I.21 presents a concrete example. The flexibility of the system arises because various production tasks can be carried out without major refitting costs, since the refitting procedures are largely integrated in the production process. The sequence of operations can also be flexibly determined, since conveyance is not based on a particular order of machine runs. Since the processing stations are provided with NC programs from the control computer of the flexible production system, this can be construed as a DNC system. The individual processing stations are in general CNC systems, but may also be more extensive processing centers.

54

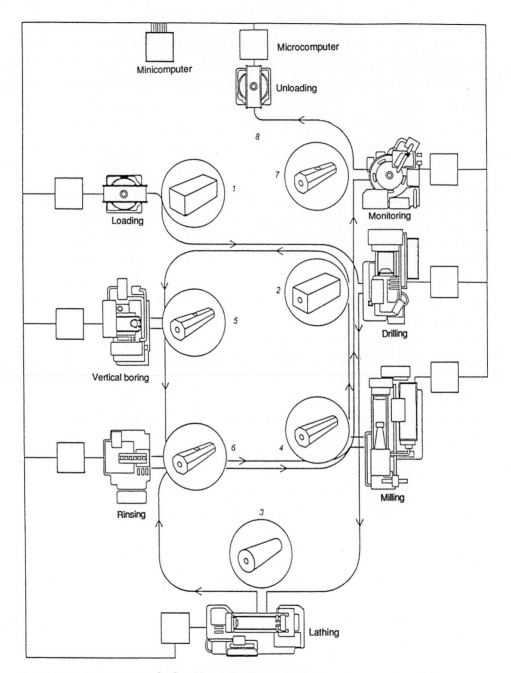

Fig. B.I.21: Components of a flexible production system
from: *Gunn, Konstruktion und Fertigung 1982, p. 95*

2.4 Production and Assembly Islands

This organizational form is less related to the use of computers, but rather is based on the criterion that production and assembly islands can carry out the processing of end products or assemblies from start to finish using given starting materials. The required equipment is established in accordance with the production flow. Although production and assembly islands can be established for manual processing forms (which may be justified by the greater process integration achievable, the consequent reduction in processing times and the improved motivation of the work force) there is nevertheless a close link with the previously mentioned computerized organization forms. Hence a production island can be organized in the form of a flexible production system in which planning or specific machine control functions, including conveyance and storage, are carried out for a specific parts spectrum by a control computer. The most important criterion for a production island is that all resources needed for production within the island are allocated to the island, and that planning and control functions are autonomously coordinated. The choice of parts suited to island production or assembly can be made with the help of statistical similarity investigations (cluster analysis). Although it is a basic principle of the island to be as autonomous as possible, this does not exclude the possibility that interactions will take place between different islands producing different parts spectra. This may arise when, after reorganization of the obviously suitable parts, other parts must be produced within the islands which do not quite correspond to the ideal requirements. It is then conceivable that parts will have to leave the island for specific processes and be reintroduced later. In this case the control system must be capable of following these order movements. This imposes considerable demands on the flexibility of the island's control system. These kinds of functions are included in a well equipped PPC system, so that once more a close relationship between the technically motivated organizational forms and the functions of production planning and control arises.

2.5 Flexible Transfer Lines

The organizational forms represented for smaller output levels also influence the production technology of large scale processes, which are generally carried out in transfer lines. A flexible transfer line aims at speedy refitting, and hence adjustment to changing production orders. At the same time the general characteristics of a transfer line are retained (adjusted material flow and precisely timed transfer of production parts within an optimized layout of processing stations). The flexibility of the transfer line depends on all of its components - conveyance, material flow and individual processing stations.

The computerized production systems that have been mentioned are often primarily considered as experimental systems and hence as isolated solutions. It can be seen, however, that they can be combined to form ever-expanding concepts, and hence demand a fundamental decision on the part of the enterprise as to which organizational form should be used for each parts spectrum, and how the production systems should be connected with one another. This demands careful layout planning. Here reference is made to a development in which, through increasing use of computers, potential manufacturing forms determine the total enterprise structure via layout planning of all components (location of production units, material flows and information flows (see *Bullinger, Warnecke, Lentes, Factory of the Future 1985, p. Il*).

e. Computer Aided Quality Assurance (CAQ)

Questions of quality assurance and control accompany the entire material flow, starting with the checking of incoming materials, the quality control of the production process itself, and the end control of the finished product. These checks are of increasing significance, since the late discovery of any mistake can lead to excessive costs. In many industrial concerns quality assurance constitutes as much as 50% of production costs. Computer support can be introduced at two levels. First, checks can be increasingly automated (analysis instruments, sensors, counters, etc.) and second, the **planning** of the checking process can be carried out in the same way as computerized production planning. Numerous test procedures have been developed by statistics and operations research for planning quality checks. In general there is a tendency towards the integration of quality checking with the production process itself. This is the case, for instance, when quality control consists of quantity or weight checks. The computerized planning of quality control relates to the support of the administrative process. Hence the proposed quality control test plan for a specific part can be seen as a kind of work schedule in the PPC sense. Independent systems can be introduced to store the checking plans, or they may constitute individual processes within the production work schedule.

f. Maintenance

As a result of increasing automation it is now often the case in industrial concerns that the majority of employees working in production are concerned with maintenance. A distinction can be drawn here between preventative and curative maintenance. Preventative maintenance involves the replacement or inspection of a machine or machine part on the basis of a maintenance plan. The optimization techniques used to determine the optimal maintenance interval, the size of the maintenance group and combinations of maintenance procedures are themes within the framework of models of preventative maintenance that have been developed within operations research (see *Scheer, Instandhaltungspolitik 1974*).

As was seen in the area of quality assurance, the planning of computerized maintenance measures can be seen as an analog of the planning of production orders. For maintenance operations which recur frequently the same procedures can be employed for process planning as are normally used in production. Closely related to this problem is that of the scheduling of the necessary spare parts. It is usual to order spare parts for a machine at the same time as the machine itself is acquired. The allocation of spare parts to machines forms a kind of bill of materials structure. Of course, in contrast with the management of production bills of materials, this does not involve the complete description of the machine, but merely of those parts held in store for the machine. As with the management of the general bills of materials, the use of this data structure can produce evidence of the use of certain parts, so that parts no longer needed can be identified. The installation of a part into a machine can be interpreted as a production order, so that the principles of production control and order tracking can be applied here too.

II. Interfaces Between CIM Components

The justification of CIM is based on the linked operational structure of the CIM components. This can be characterized by the data relationships existing between them.

a. Data Relationships Between CAD and CAM

Data flows between CAD and CAM determine: the production data requirements from the geometry of the production part, the production processes and the information content of the CAD system (see *Diedenhoven, Informationsgehalt von CAD-Daten 1985*). The

58

geometry data are taken from the data structure of the CAD system. Here they need to be converted, since not all geometric information (e.g. shading) is relevant to production.

Fig. B.II.01 shows the relationship between the control possibilities for a milling machine and the results required in the production of a production part. To a greater or lesser extent divergences from the planned form are apparent. In contrast, with 5-axes control, in which any desired tool location can be attained, divergences can ultimately only be due to the deformity of the tool. The more complex the control possibilities are, the greater is the need for complete geometric description from the CAD system.

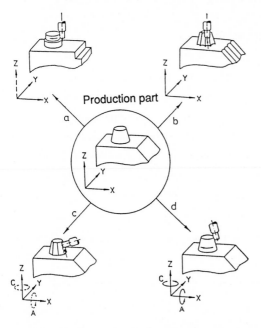

Fig. B.II.01: Alternative control systems for a given milling process
from: *Diedenhoven, Informationsgehalt von CAD-Daten 1985*

Since all manufacturing is in principle volume oriented, the complete transfer of geometry data requires 3-D CAD systems. Even for the production of flat sheet metal parts, for which a 2-dimensional plan drawing is at first sight adequate, knowledge of the third dimension - the thickness of the part - is required. For this reason 2-D systems require that the missing data be (interactively) supplied for NC control. Of the 3-D models, edge models, surface models and volume models are in this order increasingly suited to data utilization. In particular, collision experiments, which are designed to eliminate undesired contact between tools and production parts, can only be satisfactorily carried out with volume models. Of course, the higher data requirements of volume models have to be taken into account, so that for simpler production sub-tasks (e.g. determination of tool paths in only **one** processing level) a 2-D system can also be employed. For this reason the combined use of 2-D CAD systems and 3-D volume CAD systems is the most economical approach for transferring the geometry data to NC programming.

The direct transfer of geometry data to NC programming is referred to as a CAD-CAM link. Although the need for avoidance of redundant data storage and repeated data entry is acknowledged, it is by no means self-evident in all systems. The reason for this is that the producers of Computer Aided Design programs do not always offer programs for computerized manufacturing and vice versa. Great efforts are being made, however, to establish standards and norms which will facilitate the direct transfer of data from the CAD system to NC programming. In this context the so-called IGES interface should be mentioned, with whose help CAD systems can be linked together (see Section B.I.b.3 above).

The geometry data taken from the CAD system are supplemented with technological data either taken from an existing work schedule or entered interactively by hand. From this information a machine independent program code is produced (e.g. in the NC language EXAPT or APT - see Fig. B.I.19). As soon as the relevant equipment is ascertained the

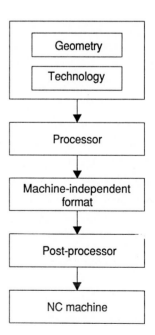

Fig. B.II.02: NC interfaces

program is adapted to the specific properties of the equipment and to the tools and materials to be used. In this way an equipment-specific program is created. The adaptation of the machine independent NC program to the control format of a specific piece of equipment can be carried out automatically using post-processors (see Fig. B.II.02). In general, however, additional intervention by programmers will also be required.

The bracketing together of CAD and CAM also aids in the automatic generation of work schedules, in that standard times can be established from the geometry and technological data using stored nomograms, etc. In addition, by using simulation programs, processing paths of the tools can be followed by the computer in accordance with the part geometry, and in this way the time values can be automatically established. The tool path can be represented visually on a terminal. Whereas work scheduling for conventional production units occurs largely within the organizational unit process planning prior to the implementation phase, the creation of NC programs involves a closer relationship with the production process. This applies particularly to the addition of technological data relating to the equipment and tools to be used at the workplace. For this reason, alongside central program creation (e.g. in the process planning department), so-called shop floor programming or "in place" programming is being pursued as an organizational alternative. The increasing computerization of production systems and just-in-time production at once gives rise to both the technical possibility of, and the need for, programming.

b. Data Relationships Between PPC and CAD/CAM

Data relationships constitute a significant interface between technical and managerial data processing. The data requirements made by the PPC system on the data produced by the CAD/CAM system must, therefore, be explored below. Thereafter, the reverse data flow from PPC to CAD/CAM will be analysed. In each case distinction must be made between primary data, customer order data and production order data.

1. Data Flow From CAD/CAM to PPC

Fig. B.II.03 shows the important data flows from CAD/CAM to PPC. At the same time alongside the PPC functions vital managerially relevant decisions are shown, which can be allocated on the basis of their degree of impact on these functions. This indicates the decisions for which CAD/CAM data can be drawn upon.

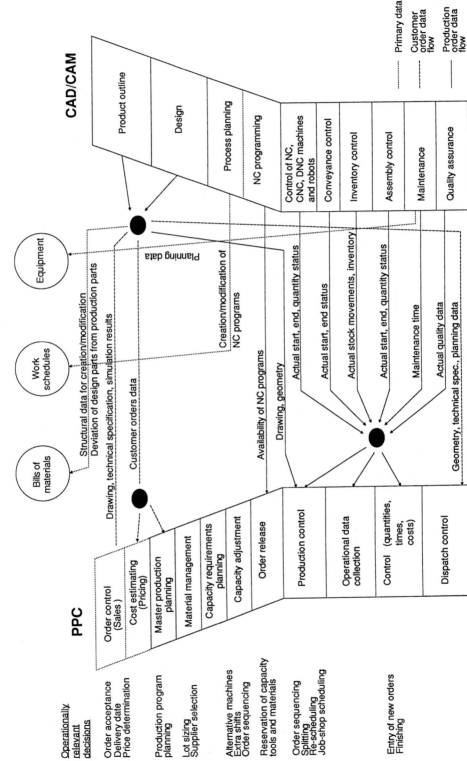

Fig. B.II.03: Data flow from CAD/CAM to PPC

1.1 Primary Data

Here we are merely concerned with the data flow from CAD/CAM to PPC primary data, but not with data distributed from the primary data to individual PPC functions, since every PPC function accesses one or more of these data types. The creation of the drawing within the CAD framework already establishes the essential information for defining a modular bill of materials. The drawing specifies the individual parts (i.e. components) of the constructed assembly. Hence, the bill of materials can be directly derived from the technical drawing for the PPC primary data. For each part defined in the drawing a parts group is specified and the number of components used in the higher level part (assembly) can be read automatically from the drawing. This automatic bill of materials generation not only diminishes the data collection effort in comparison with separate management of this data in the CAD und PPC areas, but also, and most importantly, improves data integrity via the uniformity of updating of bill of materials information in both areas. One difficulty arises, however, in that the design bill of materials produced within the CAD framework generally collects parts that in design terms belong together into a higher level assembly, while in the PPC framework a production-technology-oriented bill of materials is required, which collects components together according to the production flow. To resolve this problem, it is possible to manage both production and design bills of materials "in parallel" in a database, in that they contain part groups and common structure groups only once. Through the definition of separate structural relationships, which belong either to the production bill of materials or to the design bill of materials, further logical aspects of these primary relationships can be established.

Of course, the information contained in a CAD bill of materials will not be as extensive as that contained in the parts and structure groups. These primary data must still be prepared by the PPC area, along with information about processing times, delivery dates, costs, suppliers, etc.

In the design phase the designer has access to a library of standard geometric elements (circle, line, sphere, cuboid, ...) which can be projected onto a screen by activating a function of the graphic tablet. In addition, a library of standard parts, including norm parts, is available, which can also be copied into a drawing. Particularly important, however, is the use of existing drawings for order-related parts. For every design task the designer should establish whether an identical or similar part has not already been designed, from which the existing drawing, bill of materials description, etc., can be adopted. The avoidance of repetitive work is the primary source of the efficiency of CAD systems. Hence, their use in variant design is greater than in original design (see *Anselstetter, Nutzeffekte der Datenverarbeitung 1986, p. 83*). The search for similar parts is, therefore, crucial in judging the quality of the computer support of design processes.

This assumes that a computerized similarity search of bill of materials data is possible from the designer's workplace. The similarity search might be conducted using a classification system for the parts spectrum (a well-known example of this is the Aachener-Opitz Schlüssel, see *Opitz, Klassifizierungssystem 1966*) or similarities within the parts can be identified using statistical procedures such as cluster analysis, etc. If textual information is used on a large scale to describe parts, text retrieval systems can also be used.

In the context of computerized work scheduling, including NC programming, work schedules are produced which also access the primary data of the work schedule files of the PPC system. Where NC programs are automatically generated from CAD drawings it is also necessary to add information about set-up times, processing times, etc. to the work schedule data.

The maintenance system delivers planning data about preventative maintenance measures to be undertaken to the equipment management. These data are taken into account in the capacity scheduling and production control of the PPC system.

1.2 Customer Order Data

In customer-oriented production design activities are incorporated into the order acquisition process. For instance, it is beneficial to order acquisition if the customer can be provided with technical drawings at an early stage, and if the feasibility of the ordered technical specifications can be scrutinized. Hence CAE/CAD is particularly important for sales-oriented functions, and is taken account of in setting delivery dates. Similarly, the early availability of cost information is needed by sales for price determination. To make a rough calculation of costs and capacity requirements, rough data about the bills of materials, production methods to be used and external supply of critical parts is needed at an early stage. For products which may be difficult to transport geometry data is required by shipping to ensure the availability of adequate shipping facilities at the appropriate time.

1.3 Production Order Data

The availability of required resources is checked in the context of order release. This includes checking the availability of NC programs to be used for the production order.

Production control requires the drawings, including measurements, as production documents. The automated production, conveyance, storage, assembly control and quality assurance systems deliver actual information about achieved starting and finishing dates, as well as produced output and quality to the organizational data collection system. This data can in part be directly transferred from the production units to the data collection system. The data in this system serves as a basis for production control. They can, however, be used simultaneously for current statistical cost accounting within the accounting system, as well as for performance-related gross wage calculations. In the control context divergences between planned and actual values can be analysed, and fed back directly into the machine control adjustment.

2. Data Flow From PPC to CAD/CAM

Fig. B.II.04 shows the data flows from PPC to CAD/CAM. In the CAD/CAM area managerially-relevant decisions are made, whose consequences are conceivably not adequately clear to those responsible.

2.1 Primary Data

Fig. B.II.04 merely shows the data flows from the primary data to the CAD/CAM functions, but not the particular source of these data in individual PPC functions. Integrated CAD/CAM production takes into account at the design stage which production operations and which resources should be used. In this context we refer to production-oriented design. The higher the degree of automation within production through the use of robots and NC machines, the more precisely the characteristics of the production units, including their available tools, must already be known at the design stage. This means that primary data relating to equipment, particularly equipment and tool specification, must be accessible during the design process. At the same time capacity information for equipment must be available from primary data management and capacity requirements planning and adjustment, so that, for example, the use of bottleneck equipment can be avoided in the production of rush orders. As well as equipment data, the standard production procedures to be used for in-house parts must be available to the designer.

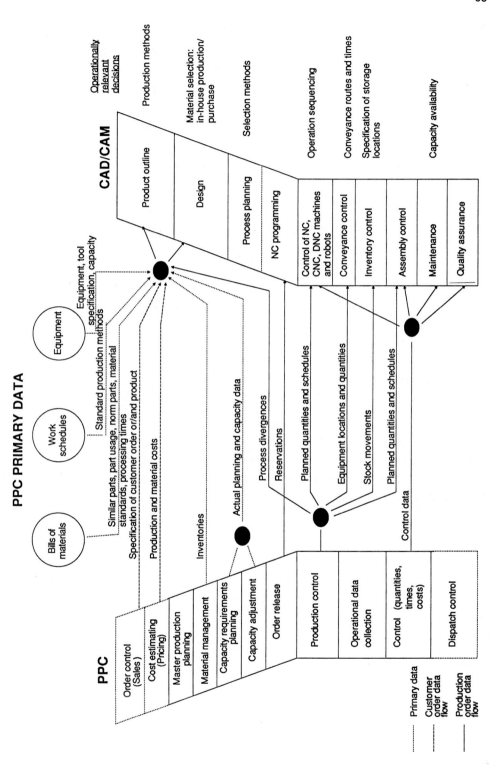

Fig. B.II.04: Data flow from PPC to CAD/CAM

Particular rationalization success is expected in the CAD/CAM area as a result of group technology (see *Teicholz, Computer Integrated Manufacturing 1984, p. 169*). This means that as far as possible design will make use of existing parts or drawings. This requires the existence of a catalog of components (bills of materials) as well as a powerful retrieval system to search for similar parts.

To allow for customers' desired delivery dates the designer also needs to have access to processing times for critical parts and supply times for purchased parts. With the use of this information it is possible for the design department to make the necessary decisions about the production procedures to be employed, the materials to be employed, and between in-house production or purchasing of parts. This can be particularly effective when the costs of alternatives (materials, production procedures, etc.) are also included. This deals with the criticism, applicable in the normal case, that the predominant proportion of production costs are determined by the design department, while the PPC functions have a very small capacity to influence costs (e.g. by lot sizing, sequencing, equipment allocation, choice of suppliers, etc.).

2.2 Customer Order Data

To produce customer-demand-oriented design (e.g. variant solutions) the specification of the customer order is required. At the same time, material costs and accounting programs from the PPC area are used to make decisions as to production procedures, etc. This requires attention not only to customer order related data, but also, for example, to the capacity situation or inventory levels of critical materials.

2.3 Production Order Data

Resources for released orders are reserved in the course of order release. This also refers to the provision of the NC programs to be used. Production control sends a multitude of control impulses in the implementation phase of Computer Aided Manufacturing. These relate on the one hand to production start-up through specification of planned deadlines and planned output levels for operations to be carried out in the course of parts production and assembly. In the course of conveyance control, production control specifies equipment locations and quantities to be transported. For inventory control, quantities to be deposited or withdrawn are specified by production control. The monitoring of the planned and actual quantities, as well as planned and actual deadlines from the operational data collection generate adjustment impulses to regulate the machines or control the system responsible for the divergence.

C. Implementation of CIM: /Y-CIM/ Information Management

I. Procedures and Project Management for Developing a CIM Strategy

Establishing uniform process chains with their accompanying data flows and short term control possibilities gives rise to great rationalization potential. This explains the substantial user-acceptance of CIM principles. What can a user do, however, when CIM is widely discussed, but not (yet) available in the form of complete CIM systems?

1. Wait until CIM hardware and software is available for full implementation.
2. Proceed with partial solutions and hope that, when CIM is fully available, these partial solutions can be integrated.
3. Make fundamental decisions relating to PPC, CAD and CAM such that they will be in line with the future CIM set-up, but otherwise proceed with partial solutions.
4. Actively pursue all the present opportunities for implementing CIM.

The first two options have the advantage that they need not involve "learning from one's mistakes" in the difficult task of CIM development, but have the disadvantages of the risk of loss of "know-how" in a future-oriented area, and dependence on the development strategies of producers of CIM systems.

The last two options require, to a greater or lesser extent, an overall concept. The development of such a concept certainly demands considerable effort, but clarifies the enterprise's position with respect to CIM and shows the direction in which the entire enterprise structure should develop.

Strictly speaking, a CIM strategy should embrace the entire future concept of the concern, beginning with the question of location, operational structure, organizational structure, the production program to be undertaken, the extent of standardization, the layout of the factories, down to manufacturing techniques and the associated information system. Hence CIM becomes a part of strategic planning. Consequently a CIM strategy is associated with **top-down planning**, as shown in Fig. C.I.01.

An important consideration is the early persuasion of the middle management. It is evident that these employees feel particularly threatened by the introduction of sophisticated new technology. Coordination functions, previously taken care of by these

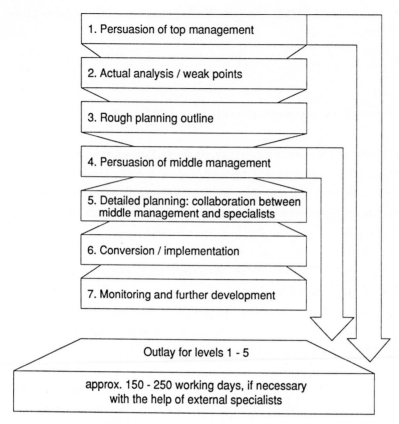

Fig. C.I.01: CIM strategy

employees, will disappear in the process of streamlining organizational processes. On the other hand supervisory personnel will experience job enrichment as a result of reintegration effects. Even when this is not explicitly stated, middle management tends to have a fine sensitivity to such possibilities, and to experience them as a threat. It is, therefore, essential that new functions for this management level should be defined at an early stage in terms of increased planning, control and development responsibilities.

Along with top-down planning, a **bottom-up strategy** is also necessary. A CIM system only has any chance of success if it is accepted by the specialized departments. For this reason, constant switching between bottom-up and top-down approaches is needed in the strategy development process: critical factors must be fully discussed before incorporation - otherwise, the ill-prepared presentation of a strategy proposal can lead to blocking of its acceptance.

A CIM strategy demands considerable organizational knowledge of the informational interdependences and processes within the industrial enterprise. Furthermore, familiarity with hardware and software developments in information technology is essential. For these reasons the employment of external specialists will often be needed for strategy development. In the author's experience the development of a CIM system for levels 1 to 5 of Fig. C.I.01 for an enterprise of about 1000 employees requires 150 - 250 project days.

The development of a CIM system can take between six months and a year, depending on the size of the enterprise. The process can be divided into two phases:

1. Phase: Investigation of the actual level of integration using process chain analysis.
2. Phase: Generation of the CIM planned integration system.

A project team, composed of specialists from the key areas, is formed to carry out the **actual analysis** (see Fig. C.I.02). To be fair to the meaning of CIM, a steering committee should be formed from the leaders of the areas concerned.

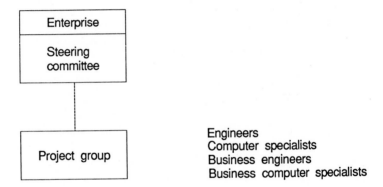

Fig. C.I.02: Project management for the actual analysis

In the first phase the project team carries out an actual analysis of present procedures from an integration standpoint. This will involve analysis not only of the computer system, but also of the manual processing sequence, so that interfaces (e.g. computer-technical) and organizational discontinuities within the operation of a homogeneous process chain can be identified (see Figs. A.II.01 and A.II.02). Typical process chains, which must be followed strictly from beginning to end are: order processing (order acceptance, design, material and capacity management, manufacturing, dispatch) and production information (drawing, work schedule, NC program, quality feedback).

The procedures are recorded via pre-constructed interviews and documented using graphical evaluation methods (process chain diagrams). The results are given in an

72

executive presentation. This presentation are addresses itself particularly to the organizational and informal discontinuities. This establishes where, in the course of order processing, significant waiting times arise which could be avoided by strict integration, reducing the order processing time scale, foɪ example, from 6 to 2 weeks. As a visual aid, the degree of integration and computer support of the individual components of a CIM architecture can be represented in the same form as the Y-diagram of Fig. A.01. The individual steps can be characterized according to the stage of development by shading or coloring, so that weak spots and omissions can be identified immediately. In the same

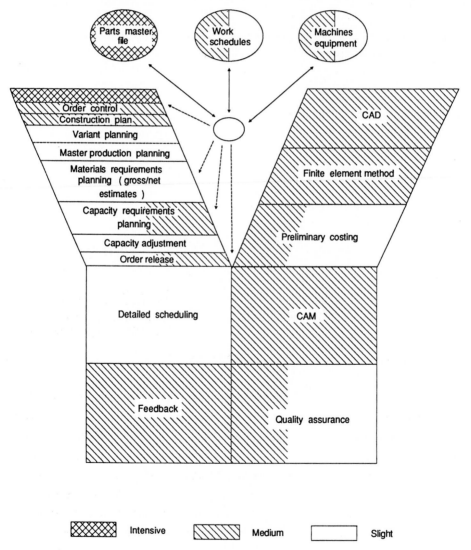

Fig. C.I.03: Data processing support in production
    from: *Lingnau, Realisierung eines CIM-Konzepts 1985*

way, in the course of developing of the planned system, the desired level of integration can be shown graphically on the Y-diagram. Fig. C.I.03 shows the starting point in the specific case of an individual automobile firm. Simultaneous presentation of the weak points of the present systems allows development priorities for the planned concept to be established.

In the course of the second phase, that is the **development of the planned system**, working parties are formed for the essential areas to be dealt with. This causes a fanning-out of the project organization (see Fig. C.I.04). Typical working parties might be:

- material management,
- interface between centralized and decentralized production control,
- integration of CAD with material management,
- integration of production control with CAM.

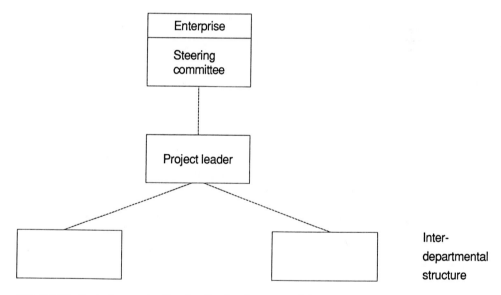

Fig. C.I.04: Project organization for the development of the planned system

The composition of the working parties should reflect the fact that there will be functional overlap, in that they should involve workers from diverse relevant departments. The working groups should consist of between three and seven members.

In the course of their work new procedures are developed for their area, which avoid the weak points of the old procedures. Finally, computer support system concepts are developed.

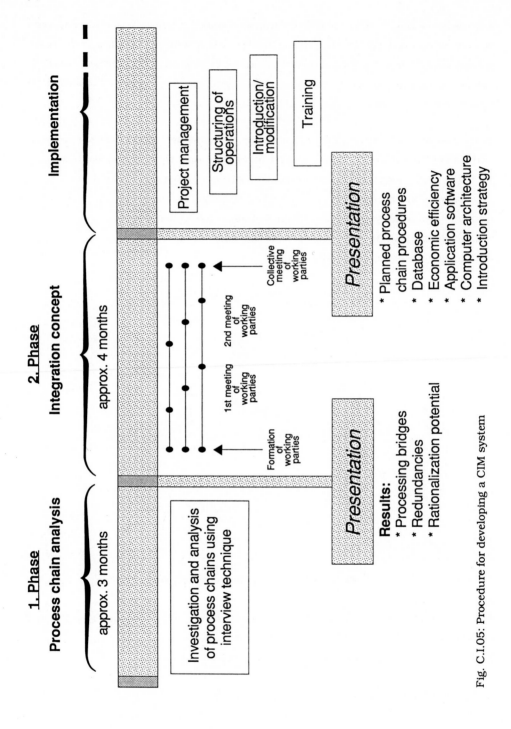

Fig. C.I.05: Procedure for developing a CIM system

In the final stage the results of the working parties are combined to form a **total system**. The entire process is shown in Fig. C.I.05. The selection process for standard software to cope with concrete problems can be established on the basis of the requirements profile. The organizational procedures also largely determine the list of requirements from any future hardware architecture, particularly its degree of decentralization.

Efficiency analysis, consisting primarily of improved time effects of integration and the consequent freeing of current asset capital, is needed to gain the support for the system from the top executives of the enterprise. An introduction plan, based on the qualitative and quantitative data processing capacities and the specialist departments, specifies the priorities and sequences for concrete system developments (see the bar diagram in Fig. C.I.06, which shows the implementation strategy for the case shown in Fig. C.I.03).

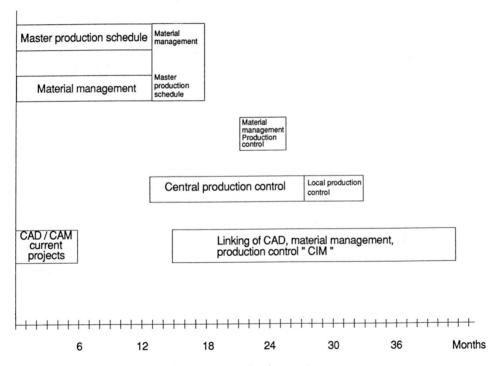

Fig.C.I.06: Timetable for a total CIM system development process

II. Establishing Objectives

The starting point for a CIM system is the strategic planning of the enterprise. For example, the enterprise can orient itself according to one of three general objectives

- cost leadership,
- product variety,
- market position.

The analysis of competitive factors and critical success factors for the enterprise can also provide reference points.

In all events fundamental considerations relating to the product program to be pursued, sales and purchasing markets, and the internal logistics and production technology of the firm are the starting point for a CIM strategy.

The graphical representation of the production profile of the firm, such as is presented in the throughput time-quantity profile of Fig. C.II.01, can provide a basis for such considerations. Quantity is represented on the horizontal axis and time on the vertical axis. Fig. C.II.01,a indicates that bought-in parts are first put into store (average time in store is indicated), then distributed in the course of parts production. These parts are then in turn stored and subsequently assembled into a smaller number of complex units. A variety of end products are then created from these units in the course of final assembly. The time for processing customer orders is also specified.

Fig. C.II.01,b indicates the firm's intended production profile for the coming years. Firstly, it shows a drastic reduction in order throughput times. Simultaneously, the number of suppliers is reduced by introducing just-in-time cooperation. Comprehensive support of variant possibilities increases product diversity and thereby extends the customer spectrum.

Given such a specification of goals the aim is now to find suitable CIM components which are capable of supporting both the temporal and the quantitative changes within the production profile. For example, the use of CAD systems could facilitate the design of variants, and the simultaneous introduction of classification systems and a greater stress on group technology could reduce the number of assemblies.

The introduction of assembly and production control systems could reduce throughput times in the assembly and production area. Other CIM components such as CAQ or DNC operation can also be analysed in similar fashion.

The compilation of a throughput time-quantity profile and the investigation of the structure of goals to be achieved are possible by means of interviews and discussions. But even a single strategic objective, such as "In the next few years we want to reduce the number of suppliers by half" can form the basis of a CIM system. In any case it is clear that the strategic definition of goals must be the starting point in the process of developing a CIM strategy.

These generalized objectives are then rendered more operational in the subsequent evaluation of concrete CIM sub-projects (process chains).

77

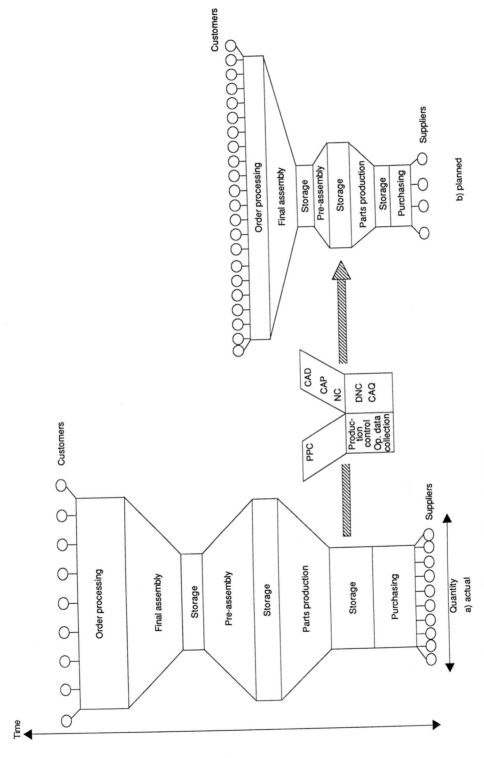

Fig. C.II.01: Throughput time-quantity profile DUMP

III. Definition of CIM Process Chains

Since a complete CIM system is still not available for sale, a user who wishes to implement CIM can only start with CIM sub-chains once the necessary framework has been defined. The choice of sub-chains depends on branch- and factory-specific

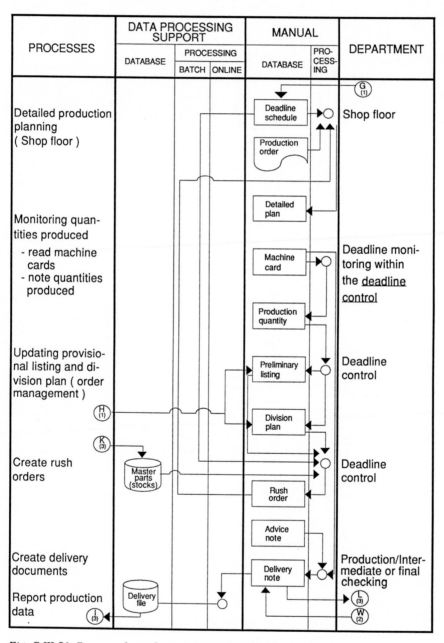

Fig. C.III.01: Process chain diagram: Actual production control process

characteristics. Typical CIM sub-chains are first highlighted and then factors influencing the priority to be attached to their implementation are discussed.

The significance of the individual chains depends on the branch of the enterprise and factory-specific factors.

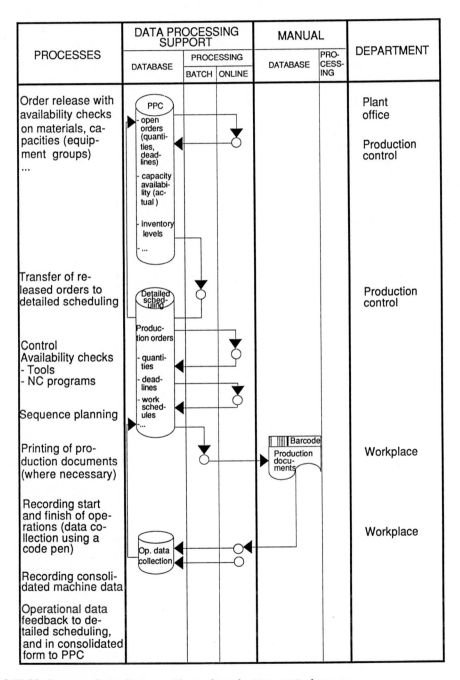

Fig. C.III.02: Process chain diagram: Planned production control process

Process chain diagrams can be used to analyse the present organization of process chains and to develop the planned model.

Fig. C.III.01 indicates the actual state of the process chain for production control. It shows that the processing switches several times between computerized systems and manual procedures. These organizational "discontinuities" frequently give rise to data redundancy and time delays. The diagram, which is largely self-explanatory, also indicates the departments responsible for carrying out the individual processing steps. This explicitly follows the notion of putting the homogeneous process, rather than department-specific procedures, in the foreground.

Fig. C.III.02 shows the formulation of the new planned process. Here it is clear that the discontinuities between manual and computerized processing are replaced by a uniform, largely on-line system.

The CIM sub-chains to be analysed below are combined in Fig. C.III.03.

81

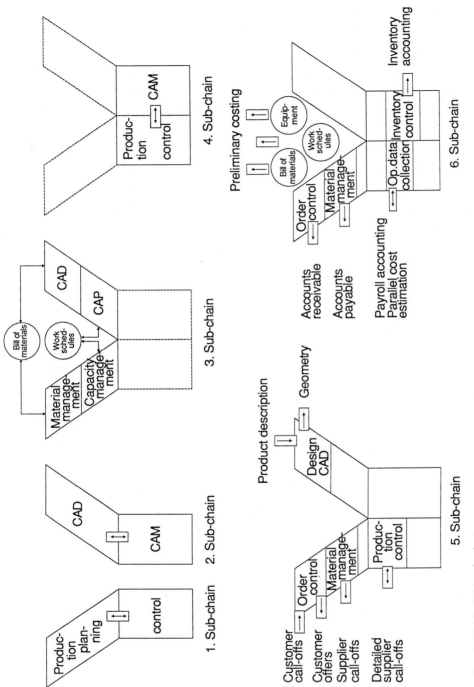

Fig. C.III.03: CIM sub-chains

a. Sub-Chain 1: Linking Planning and Control

As was made clear in the description of current PPC systems, customer order processing and the medium term planning levels of material and capacity management are dominant. Long term master planning and short term control functions are comparatively unsupported. The great emphasis placed on customer order recording, material and capacity management have also led to the situation in which PPC systems have been chosen and introduced primarily on the basis of the proximity of these sub-areas with the associated commercial functions. For example, there are close relationships between customer order handling and accounts receivable, purchasing and accounts payable, and between master data management (bills of materials, work schedules) and cost accounting.

These medium term planning functions with close links to the commercial accounting functions are accompanied by a "centralized" planning philosophy, in which a unified PPC system supports all planning and control functions down to equipment operation time (see Fig. C.III.04,a).

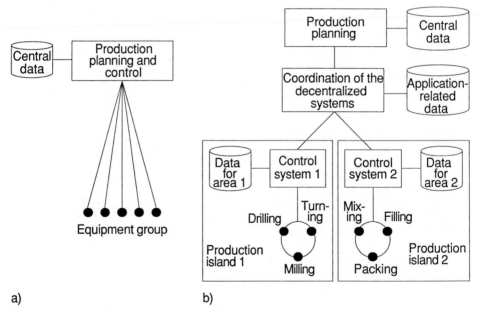

a)   b)

Fig.C.III.04: Control systems

The susceptibility of short term control functions to unplanned events (machine breakdown, employee absence, material defects, etc.) renders such a centralized model of doubtful worth. In addition, the demands regarding hardware availability, response time characteristics, and the compatibility of diverse peripheral systems also constitute significant problems for a host-oriented system. The problems that are already becoming apparent from these arguments only failed to be recognized early on because this kind of concept had scarcely been comprehensively implemented at that stage, as has already been indicated in the upper part of Fig. B.I.07.

The inadequacies of classical PPC systems in handling new control philosophies have led, for example, to situations in which the introduction of Kanban control systems has caused the breakdown of the existing PPC system. An interface for "unplanned withdrawals", which was only intended for exceptional cases, is regularly used for stock withdrawals within an autonomously controlled Kanban system. The PPC system's requirement breakdown is only used to determine the requirement for bought-in parts, determination of production orders and the entire stock-keeping position is taken over by an autonomously controlled Kanban system (see *Scheer, CIM in den USA 1988, p. 41 ff.*).

Currently observed trends in the production area, therefore, point in another direction. The development principles explained above relating to production decentralization on an object basis by the formation of production islands, processing centers, flexible production systems, etc, give rise to independent planning areas. These make different demands on production control. For example, the industrial firm generally distinguishes between parts production and assembly. Furthermore, diverse issues of a production-technical nature can play a role in control: If machine utilization is dominant, sequencing problems relating to the reduction of refitting times is an important control criterion. If valuable materials are being used, layout optimization for diverse orders to avoid wastage and rejects will be the decisive control principle. The formation of optimal kiln batches might necessitate yet other criteria for combining orders into planning units (lots, sequences).

As a consequence, an order as defined in the material management context, which always relates to a parts number, need not necessarily be the relevant planning unit in the production control context: instead, combining and splitting can generate new planning units at the operation level.

This state of affairs is expressed in Fig. C.III.04,b by the formation of independent planning areas. Since these must be supplied by a higher level order creation system, and since disruptions in one area can have consequences for other areas, a coordination system needs to be superimposed on the decentralized control units.

Thus, the overall implication is a future requirement profile for production planning and control as indicated in Fig. C.III.05.

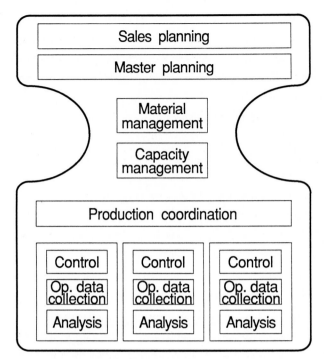

Fig. C.III.05 : New PPC system architecture

Longer term capacity management issues are taken over by a master planning system with simulation and optimization support. At the same time short term control functions are increasingly incorporated in the decentralized control systems of the autonomous sub-areas, on which a coordination system is superimposed.

The first steps towards this structure can already be seen in new application software developments. The construction of a new hardware level at the production control level is supported by the so-called control center concept. Here orders generated by a central PPC system are taken over at the PC or workstation level and re-planned within prescribed time limits. In the process diverse optimization rules for combining and splitting operations are employed. The relocation of these functions to a new hardware level increases their accessibility, since the orders are now available regardless of the host computer's shifts (e.g. during the night and at weekends).

Because it uses modern user-oriented hardware developments the control center concept also opens up new types of user interface. These might be window techniques, graphical representation and colour support, as well as mouse controls. Fig. C.III.06 shows the development of user friendly output listings from the primitive graphics of centrally-oriented PPC systems up through currently available graphical control center interfaces to the photo-realistic representations of production systems being developed using simulation and animation techniques.

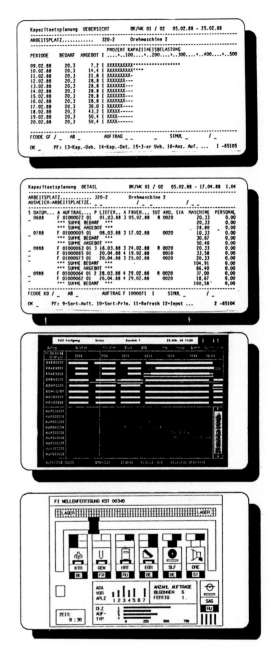

Fig. C.III.06: The development of graphical tools

A mediating level between the order-related logistical concept and technical machine control level can also be generated with the help of the control center concept. It combines data from the reporting systems of data collection functions, storage, conveyance and DNC operation and simultaneously initiates actions relating to release and control impulses at this level (see Fig. C.III.07).

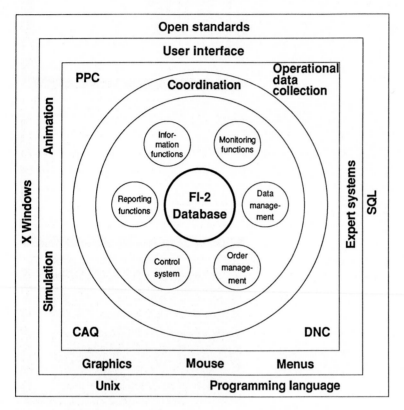

Fig. C.III.07: FI-2's open architecture

The control centers can be implemented for the control of individual planning areas. The coordination of several control centers can be supported by a coordinating control center (see Fig. C.III.08).

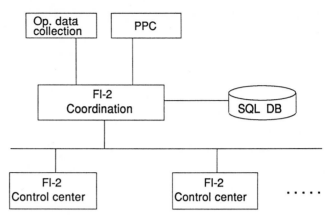

Fig. C.III.08: FI-2's coordination architecture

b. Sub-Chain 2: Linking CAD and CAM

The typical connecting links between CAD and CAM have already been shown in Fig. B.II.02. First of all the CAD system must possess an interface through which geometric information relevant to the CAM area can be passed on. For example, shading information, which can be of value in a drawing in the design area, is useless for production. Hence a processor is required, which can translate the transferred geometric information into a NC language (e.g. EXAPT). The result is a so-called machine neutral data format (CLDATA). Since the NC manufacturing machines (drilling tools, milling machines) are already equipped with control by their producers (Siemens, Bosch, Allen Bradley, Philips, General Electric, etc.), these control instructions must be adapted to the characteristics and formats of the specific machine control. This is carried out by a post-processor. These linkages must, therefore, be taken into account in the choice of CAD and NC programming systems.

The transmission of the geometry data from the design area to production is not only of interest for NC programming, but for all functions subsequent to design that process product definitions, in particular, the creation of job sequences in the process planning context. Here it may be necessary to carry out operations, such as geometric rotations, in order to specify a production process and the associated technological requirements as regards precision, shut-down speeds, etc.

If an enterprise uses several different CAD systems in the design context (e.g. electronic engineering, 2-D mechanical engineering and 3-D mechanical engineering) the information generated must be used together in the production planning context, if, for

88

example, an electronic part has to be fitted in a body that has been designed using a 3-D mechanical CAD system.

This kind of process chain, in which the geometry provides starting information for further processing steps in the production planning context, already demands a complex computer-technical integration environment.

Fig. C.III.09 shows the support for a process chain from the creation of geometry data within product development, through processing and extension of the data in the area of work schedule creation and NC programming, to completion of manufacture in DEC's CAD/CAM Technology Center in Chelmsford (Mass.).

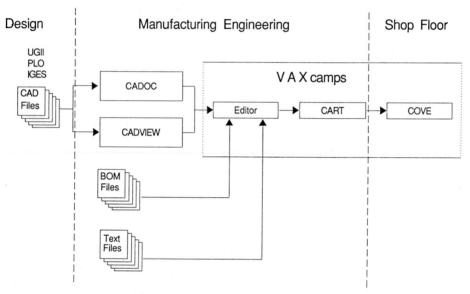

Fig.C. III.09: Process chain for product development
    Source: *DEC*

The VAXCADOC components can read diverse CAD files (e.g. from UGII, PLO or IGES) and intersperse them with text. The output can be shown on graphics terminals and workstations. The CADVIEW component makes available various possible data manipulations (rotation, analysis, removals, additions) for defining user displays and for sending data to other workstations.

While the CADOC and CADVIEW systems achieve a transition from the CAD area to further processing of production data, the components of the VAXCAMPS system support the appropriate handling and processing of data for production purposes through to completion of manufacture. An editor links geometry and bill of materials data and extends them with textual information. The CART (Camps Automated Release and

Tracking) system automates the documentation flow, i.e. the distribution, release and acknowledgement of production, test and interruption information. Using VAXCAMPS COVE (Camps Operator Viewing Environment) production is provided with access via the appropriate terminals to all the necessary textual, graphical and product history information.

Integration is achieved by the use of computer tools, such as window techniques and networking. The applications systems themselves remain largely independent.

The data flow in Fig. C.III.09 is only resolved uni-directionally from left to right. This means that subsequent changes to pre-existing files are not implemented; for example, changes to the geometry which are made on the basis of production-technical requirements within the manufacturing system, are not stored in the CAD databases of the prior design phase. Of course, the use of electronic mail systems can allow organizational measures to be undertaken to ensure that all levels affected are notified of the changes made, so that they can be followed up from there. In this way organizational data integrity is ensured, even though it is not provided in system-technical terms.

The integration flow shown here - that is, the transfer of the geometry from design to production - is merely **one** step linking design and production. That design should also take production data into consideration is of increasing importance.

The division of tasks between design and process planning and the reconciliation procedures between them become increasingly problematic with increased production automation. Therefore, a designer should take account of production-technical possibilities in the course of product development, that is, the design should be suited for production. In concrete terms, this means, for example, that in the design of a circuit board the distance between the components demanded by the automatic feeder or feed robot is adhered to. In mechanical production, too, tool properties, specification of robot and production tolerances in their effects on materials and procedures increasingly need to be taken into account by design.

This means that design must have access to data from the CAM area.

This kind of CAD/CAM integration - effective in both directions -is shown in Fig. C.III.10. It has been implemented by Hewlett Packard in their Lake Stevens works in which medical measuring instruments are produced.

Components, which are either procured externally or manufactured internally, such as semi-conductors, transistors, etc, are stored in a relational component database. These component descriptions are used as a source of information by research and development in the design of new circuit boards. The results of the circuit board design

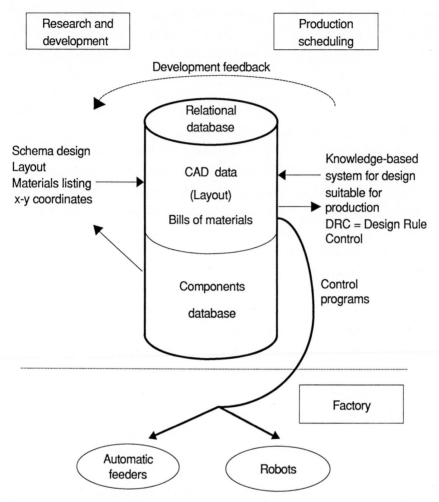

Fig. C.III.10: Knowledge based information systems for CAD/CAM integration

are also stored in a relational database in the form of CAD data, which describe the positions of components on a board, as well as a materials listing.

This design is immediately checked for production suitability by a knowledge based system. This knowledge based system contains experience and heuristic rules from production regarding the suitability of specific components and production facilities. This ensures the close integration of production knowledge and design, and avoids reconciliation cycles.

Each circuit board contains about 200 components which may be produced on different equipment by different production methods. The large number of degrees of freedom means that even at the design stage optimizations regarding production quality, production time and costs can be carried out.

The Design Rule Control (DRC) system contains details concerning the suitability of specific components as regards quality, production time and costs. In addition, the system also contains simulation procedures to discover production collisions and avoid them by suggesting more suitable components.

The DRC system thereby constitutes the link between development and production in the factory. At the same time it melts together the typically disjoint organizational areas of development and production scheduling.

This paves the way for the possibility not only of achieving system-technical links between development and production scheduling, but also of linking the two areas more closely in an organizational sense.

c. Sub-Chain 3: Linking Master Data Management (Product Description Database)

The links between CAD and CAM become clearer when the complete description of a product is regarded as a homogeneous process.

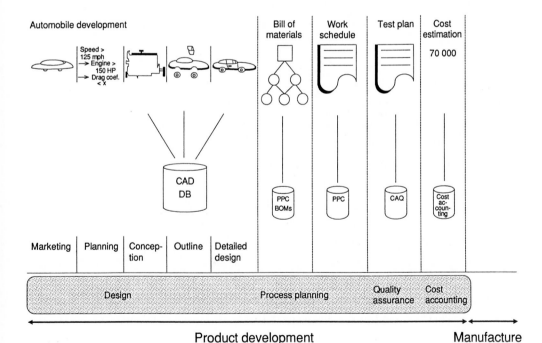

Fig. C.III.11: Departmentally specialized product development

Master data management refers to the establishing and maintenance of the master data required by the CIM components. A large proportion of these data relates to the description of the product from various viewpoints.

Product development in an industrial firm is currently characterized by a strongly specialized procedure (see Fig. C.III.11). From an original product idea generated by marketing the design department first creates primarily the geometry of the future product. In this process the geometry is increasingly refined through the planning, conception, development and finishing phases.

Once the design has been released the bill of materials is created by process planning on the basis of production criteria. This bill of materials can diverge widely from the design bill of materials established at the design stage: design thinks in terms of functional units (e.g. a car's entire hydraulic system or a product's complete electronics), whereas the manufacture of a product is structured according to the individual production steps. The production bill of materials is, therefore, also the basis for the quantity- and time-based planning within a production planning and control system.

In the next step capacity management generates the work schedule for the product. Here the individual production steps to be carried out are defined and assigned to the technical procedures and equipment to be employed. This information serves as starting data for the capacity scheduling of a PPC system.

The quality assurance department then establishes the quality attributes to be measured and develops test plans for them.

Preliminary costing then uses the bills of materials and work schedules as the basis for determining expected manufacturing costs of the product.

These data are at present managed by various systems. Since the data are generated alongside the development of the product a unified view of product description data is bound up with a unified view of the development process. On the basis of the competitive grounds for reducing development time and economic grounds (design-stage cost estimation) this process chain is increasingly of interest within CIM.

Procedures based on departmental specialization have led to undesirable consequences regarding the operational organization and computer support found in the departments. The sequential processing of the individual steps gives rise to considerable transfer times between the individual phases and their respective departments. At the same time cycles tend to arise between the processing steps: for example, process planning can declare a design as not suitable for production and thereby cause changes to be made in the design department.

The same can arise from the quality assurance viewpoint. Cost accounting can also establish that the market cannot bear the estimated production costs and thus demand changes to the product.

In addition to transfer times between the departments, therefore, the multitudinous cycles within the development chain can lead to temporal delays and increases in development costs.

Departmental processing specialization has also resulted in the use of diverse data processing systems in the different areas:

Design uses CAD/CAE systems, which manage the geometry data. The production bill of materials, which is described by a complex data structure, is recorded within the PPC system by specialized bill of materials management systems.

Work schedules also constitute part of the master data management of a PPC system.

Quality information is generally managed by special quality assurance systems (CAQ), even though a test plan shows great similarity to a production work schedule.

The data processing systems for preliminary costing also often maintain their own databases.

Although the diverse functional viewpoints all relate to a common object, they are concerned with different stages of development. Thus, bill of materials and work schedule management systems within a PPC system are only interested in the data from the planning viewpoint. This means that they only record the complete bills of materials and work schedules after release of the finished product, because only these are processed by many operative PPC systems for material and capacity planning. Although structural information is already available even for segments of the product during product development, in general these are not taken up by a bill of materials management system (exceptions are PPC systems designed for plant construction and one-off production).

Operational organization problems associated with multiple transfer times and processing cycles are, therefore, not avoided by computerization, rather, they tend to increase, since now the technical problems of data transfer between diverse data processing systems also need to be resolved.

The situation described is increasingly being recognized as presenting scope for rationalization and competitive improvements. Of course this requires the transition from specialized to parallel or simultaneous product development.

In many industrial sectors development time is becoming a considerable competitive factor. Empirical investigations in the electronics industry have indicated that introducing a product 6 months later than the competitors can imply a reduction in profits over the total life of the product of the order of 30%. In the light of this, high production costs or excessive development costs diminish in importance. The so-called "time to market" becomes the decisive competitive factor.

The reduction of development time is also of increasing importance in other sectors. The automobile industry is currently investigating how the development phase for a new car, which has up to now been of the order of 5 - 7 years, can be reduced to 3 - 4 years, since other countries such as Japan have already achieved this kind of development time.

In accordance with the findings of the network planning technique an effective reduction of process time can be achieved by carrying out sequential tasks in parallel (see Fig. C.III.12).

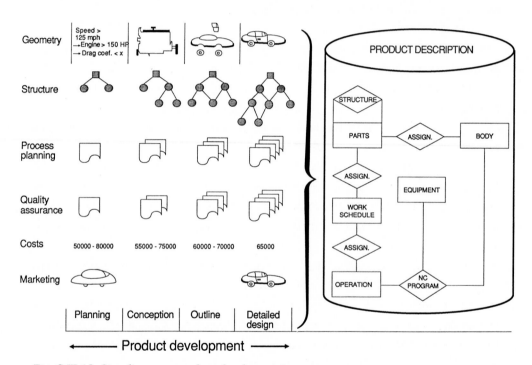

Fig. C.III.12: Simultaneous product development

This superficially simple and plausible finding is, however, exceedingly difficult to implement in practice. It requires not only a new organizational linking of the design and development tasks with the other departments concerned, but also a new philosophy of computer support.

In addition to reducing development times linking the processes in content terms can also achieve further organizational and economic effects.

The advantages of closer links between process planning and design have already been mentioned in the discussion of the CAD/CAM process chain.

A further link is that between material management (bill of materials management) and design.

During the development of the geometry the structure of the product is also implicitly established: assemblies or parts are defined during the creation of the drawing. Where the development process is very lengthy this can mean that components may be incorporated in a new design which, on release of the developed product, are no longer actively held by the firm, but which have in the interim been replaced by new parts. Thus data inconsistencies between design and material management can arise.

For this reason alone it is obvious that closer integration between design and parts management is necessary.

The early specification of a rudimentary bill of materials means that materials can already be planned. For example, for a customer-related new development with a specified delivery date a rudimentary bill of materials can allow the ordering of parts with long procurement times. This is completely normal in the project and one-off production context; here ordering processes are often carried out in parallel with design processes, it can even arise that a bill of materials is only fully detailed after delivery of the piece of plant.

The continuous, current links between the state of the design of a part and the bill of materials information in material management means that the designer can take greater account of delivery times and quality attributes of bought-in parts when making his design.

The fact has already been quoted that a high proportion of the production costs for a product are already determined at the design stage: here, requirements regarding the components and materials to be employed are determined, tolerances define the potential production procedures, and decisions as to whether parts should be bought-in or produced in-house are made. Traditional cost estimation procedures require complete work schedules and bill of materials data and thus design decisions can hardly be influenced on the basis of costs (at most only in the context of so-called value analysis, where in fact only the already existing mistakes are supposed to be corrected). In contrast, design stage cost estimation, in which all the available geometry data, bill of materials and work schedule data can be continuously evaluated at each stage of development, allows the designer to take cost factors into account.

By choosing an appropriate product form (e.g. simple accessibility of components subject to wear and tear) it is possible to have a considerable effect on the ease of carrying out quality checks and maintenance on a product. Taking such factors into account at an early stage in the design context is, therefore, also an effective requirement.

It is often the case that during the development time for a product new findings relating to competing products or changes in customer views become known, which demand changes to the product. For this reason the technical development process should be constantly brought into line with marketing findings.

These factors imply that the entire development process for a product should be organized as a unified process using the various viewpoints of geometry, structure, production, costs, quality and marketing. This requirement also implies the need to create not only a chained operational organization but also unified data management. The data overlords of the individual functions, who each manage only an subset of the product definition, must give way to a unified product description or engineering database (see Fig. C.III.12).

A unified product development database would manage the product description data from the various viewpoints using a unified logic. This need not necessarily imply unified physical data storage, but first a unified data structure must be developed. This indicates the logical relationships between parts, structure, work schedule, geometry, quality and cost information.

All the relevant functions can access this data structure, as is shown schematically in Fig. C.III.12.

In addition to the data structure the operational logic must also be supported. This means, for example, that changes carried out in one sub-area are automatically made known to the other sub-areas and initiate the appropriate actions. Suppose, for example, that the designer changes a product, this can also lead to changes in material management (e.g. in replacing one component with a new one) such that planning needs to react to the altered situation. At the same time this can also result in the need to create a new or amended work schedule for production. Quality assurance can also be affected by this measure. Correspondingly, the causative design change must immediately send the relevant trigger signals to the other data processing systems in order to initiate stoppages (e.g. to a NC program) or other activities. This kind of system support also facilitates the necessary changes to the operational organization of the enterprise.

It is recognized that close integration between design and process planning can lead to considerable difficulties and human friction, even though both carry out technical functions. Such problems become more serious when "alien" areas such as cost accounting and marketing are to be included. Considerable efforts are therefore necessary to generate the awareness of the economic advantages, that in spite of the demarcations that have arisen between the functions, ultimately **one** object is being processed, namely the product.

A unified support for product development has far-reaching implications for the structure of data processing support. Management of bills of materials and work schedules is removed from PPC systems, which thereby lose a large part of their present significance. Production planning and control then becomes simply one of several users and no longer "owner" of the data. Taking this along with the previously expressed idea of a reorientation of PPC in the direction of production control, a dramatic change in the significance and functions of PPC becomes clear.

d. Sub-Chain 4: Linking Production Control and CAM

Short term production control will become increasingly closely linked with the technical systems from the CAM area (see Fig. C.III.13). In releasing operations, production control provides the impulse to the technical systems and thereby initiates the preparation of the CAQ system test plan needed for the operation, the DNC system NC programs required, transport to be organized by specification of the location of the assigned machine, and stock movements for production parts and tools by linking the operation with tool and part information. At the same time maintenance orders are also managed and initiated by production control.

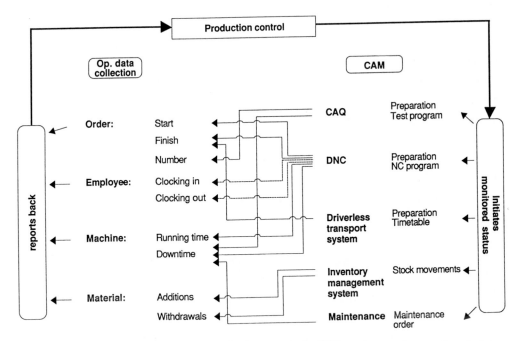

Fig. C.III.13: Data flow between production control, CAM system and operational data collection

Production control can only fulfil its short term tasks adequately if it has access to an up-to-date database of events in the firm. For this reason it is closely linked with the operational data collection system. However, if responses are only stored via terminal entry at a data entry office installed at the shop floor level, then the data will not be adequately up-to-date. The acquisition procedure is protracted by recording at the machine through to entry at a terminal in the data entry office. So it is not "actual" values that are being recorded, but rather "past" values.

Currency can be improved if feedback information is closely linked from its point of origin with production control. Simultaneously, redundancies and reconciliation procedures are eliminated in comparison with the multiple recording via manually completed response forms and terminal entry. In concrete terms, this means that the CAM components which are equipped with their own data processing intelligence, such as CAQ, DNC, driverless transport systems, and stock management systems can pass on signals directly to the data collection system.

If the CAM and data collection systems are linked then the data collection system performs a filter function between the plant control level and production control.

On the other hand, if the CAM system can also prepare (consolidate) data in the form in which it is required by production control, wage recording, or cost accounting, then a direct link between these systems and production control is possible without a detour via the data collection software system.

On the left hand side of Fig. C.III.13 some typical operational data collection like data for orders, employees, machines and material are shown. At the same time it is indicated from which CAM system these data can be acquired. For example, the start and end of an operation can be reported directly by the machine control. The end of an operation can, however, also be reported by the driverless transport system which has picked up the processed part in order to transport it to the next workplace. The number of completed parts can be reported by the CAQ system which carried out the good/bad tests.

Employee information concerning their "coming" and "going" can also be reported by machine messages, if, for example, the presence of an employee can be identified with the switching on and off of a machine. Because the correspondence is perhaps not so perfect, this link is indicated merely by a broken line.

Machine information concerning running times and down times can be recorded from DNC/SPS information.

Where the monitoring of tolerances and wear and tear on tools is increasingly integrated with production by the use of measuring and testing equipment expected and planned down time can be reported by the CAQ system.

Material additions and withdrawals can be taken directly from the stock movements of the stock management system.

The consequence of this for the implementation of a CIM system in the short term production area is that only an integrated consideration of the CAM components, production control and operational data collection can fulfil the integration requirements. This means that a careful analysis must be undertaken not only of the required type of data linkages, but also of their temporal currency, and that an integration strategy regarding database, networking, and operation system must be chosen to ensure that the necessary down-load, up-load and message transfer functions can be carried out on an acceptable time scale (see the discussion of data structures and integration tools below). The language developed for the production area in the MAP standards context, MMS/RS-511 (MMS = Manufacturing Message Specification) is of particular significance. This makes available standardized functions for down-load and up-load of programs, queries regarding device status, error messages and the call-up of program functions in device-independent form.

e. Sub-Chain 5: Inter-Company Process Chains

CIM encompasses not only internal integration within the enterprise, but also the integration of process chains beyond company limits to customers and suppliers (see Fig. C.III.14).

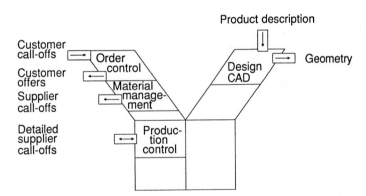

Fig. C.III.14: Sub-chain 5: Inter-company data exchange

For simple data transfer functions the electronic mail service of commercial network systems, the Videotex system of the German Bundespost, manufacturers' network systems (e.g. SNA from IBM, DECNET from DEC, TRANSDATA from Siemens) and in-house created networks are available. This file transfer allows data from one computer system to be moved to the database of another computer system (see Fig. C.III.15).

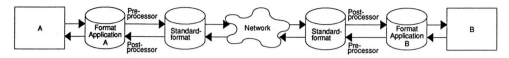

Fig. C.III.15: File transfer

Such a function presumes that standards for the transfer format are established, as shown in Fig. C.III.16 (see *Mertens, Zwischenbetriebliche Integration 1985*). In this context efforts are being made at the national and the international level (DIN, EDIFACT, etc.). The necessary reformatting from the application format to the standard format and vice versa is carried out by so-called pre- and post-processors.

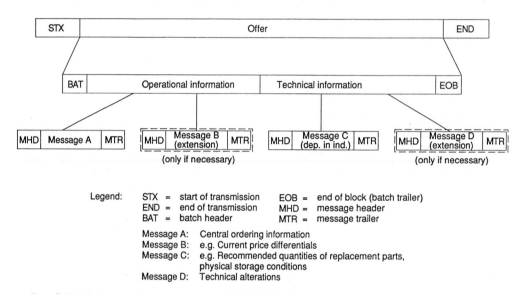

Fig. C.III.16: Protocol structure for order transmission

In an inter-application contact (see Fig. C.III.17) the programs from the various applications directly access data and programs of the other communication partner. This means, for example, that a transaction of application B can be called up by application A, which accesses data from application B and makes the results of the transaction available to program A.

Inter-application integration makes considerable demands on the standardization not only of the network, but also of the operating systems and database systems. Inter-company process integration has considerable economic potential due to:

-    temporal streamlining of processes,

-    avoidance of cumbersome paper communication,

- detailed data transfer possibilities,
- takeover of functions by the other partner or avoidance of repeated execution of the same functions,
- avoidance of planning functions via direct processing.

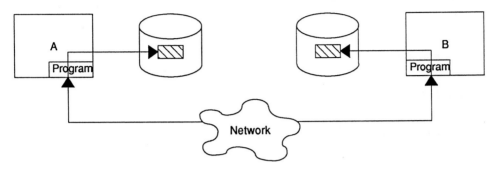

Fig. C.III.17: Inter-application contact

The last point refers to the fact that, in traditional processing, external transactions generated by one partner must first be formulated as new planning entities. With closer information chaining, however, a case-specific "just-in-time" planning principle can be applied, in which repeated batching of the results of previous operative functions can be eliminated. This processing approach, represented in Fig. C.III.18, also pertains internally, so that there is a tendency for planning functions to be thinned out prior to action.

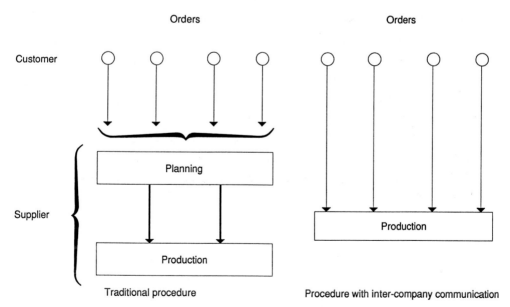

Fig. C.III.18: Transformation of external transactions in traditional processing into new planning entities

Among other things, this relates to:

- purchase planning **before** ordering,
- dispatch planning **before** dispatch,
- payment planning **before** payment,
- material planning **before** assembly.

1. PPC Integration

1.1 Initial Position

Fig. C.III.19 represents the steps in a logistic chain with the functional sequence purchasing, order processing, production planning and control, dispatch, receipt of goods, invoicing and payment handling. A three-level process between retail customer, manufacturer and supplier is considered. Reference will be made to the manufacturer-supplier relationships in the automobile industry. Each point indicates which of the three is primarily responsible for carrying out the processing function concerned. The last row shows the total number of functions carried out by each of the individual logistic partners. It is assumed that all information exchanges take place in writing.

It is clear that the process chains between retail customer and manufacturer and between manufacturer and supplier run parallel, even if a customer's considerations as to whether to acquire a car (denoted as purchase planning) are quite different to the formal purchase planning procedures undertaken within an industrial firm. Although the information flow is represented very coarsely, five documents must be passed from buyers to sellers and five documents from sellers to buyers. The physical transfer of the product is also shown (a car between manufacturer and customer, and tyres between supplier and manufacturer). The link between sales and purchasing logistics is represented for the manufacturer by master scheduling, in which the need for in-house and bought-in parts is recognized and then processed further by purchasing.

By using information technology, with overlapping data flows and the shifting or elimination of functions in the logistic chain, substantial streamlining of the logistic chain can be achieved.

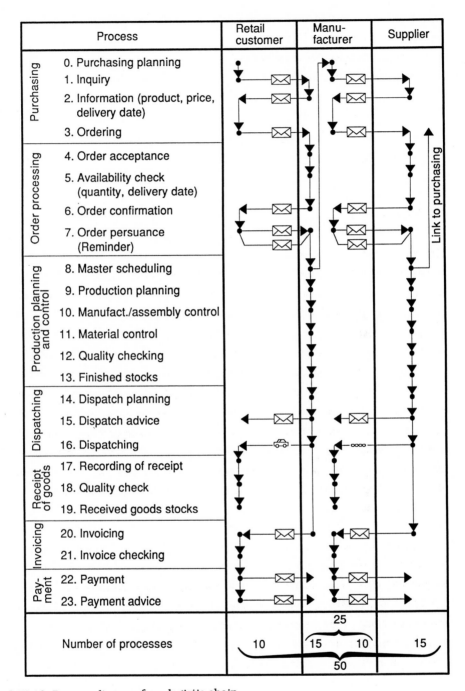

Fig. C.III.19: Process diagram for a logistic chain

First it will be assumed that simple communications between the logistic partners is possible in the form of file transfers.

Technical possibilities that could be employed are:

- electronic mail service via Videotex or other network systems,
- terminal access to data records of the other partner in the form of information collection.

Direct communication between the partners does not occur, however. A direct linkage of the application systems will be examined next.

1.2 Data Exchange

The process chains are shown in Fig. C.III.20. Alterations caused by transfer or elimination of points, each of which represent one processing step, are obvious. The customer's ability to access the manufacturer's data records, e.g. via Videotex, means that postal communication to make or respond to enquiries is eliminated, since these can be carried out directly by the retail customer. Even ordering can take place via the mailbox system. Hence, the retail customer takes over the order recording function, since, after electronic transfer, the data can be taken over directly by the manufacturer's system and subjected to availability checks. It is assumed that order confirmation continues to be effected in writing (although this is not absolutely essential). Order pursuance, i.e. queries concerning the delivery date, can also be made using the mailbox system. A system analogous to the running total procedure can be presupposed. This means that basic agreements between producer and supplier are generated, which are implemented by current call-up. These call-ups are made electronically.

As a result of the closer informational interdependence between the manufacturer and his supplier, the processing chain for the task of assembly control must be reformulated. Notification of requirements no longer comes from master scheduling, but from the subsequent area of assembly control which works with differentiated data. In the traditional processing approach, establishing the sequence in which materials are processed along with assembly control are significant planning steps for the manufacturer. The direct electronic exchange of data also allows a greater degree of differentiation. This means that for purchase planning, and hence ordering, not only quantities and dates are given, but also the sequence in which the components required for the assembly plan need to be delivered. An illustrative example of this is that for BMW in Dingolfing the supplier must ensure that the delivery of upholstery by color, etc, takes place according to the sequence of the assembly plan on an accurate hourly basis (see

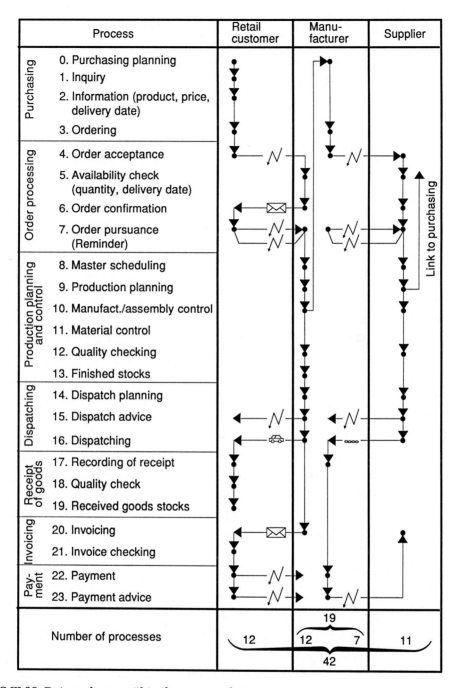

Fig. C.III.20: Data exchange within the process chain

Zeilinger, Just-in-time und DFÜ bei BMW 1986). In Fig. C.III.20 the "material control" process is transferred from the manufacturer to the supplier (to his dispatch control). At the same time purchase planning is eliminated as an independent process, since it is combined with assembly control.

The process chain between retail customer and manufacturer continues for the manufacturer with the quality control of the manufactured products. Dispatch advice can also take place electronically using the mailbox system. Payment can be effected automatically by the retail customer via a connected Videotex system with Home Banking, and, similarly, payment advice can be communicated via the manufacturer's mailbox system.

The process chains between retail customer and manufacturer and between manufacturer and supplier no longer display the same degree of synchrony when file transfers are employed as was the case with manual transfer. The closer links between the supplier and the planning system of the manufacturer allow accurate timing of deliveries in accord with the concept of just-in-time production. To achieve this the supplier's quality control must be intensified, while for the manufacturer quality control and stocks of received goods are eliminated. The supplier takes over the manufacturer's material planning and incorporates it into his dispatch planning. Dispatch advice is carried out electronically. Once the goods have been received the manufacturer takes over responsibility for payment, without having received a formal invoice. Since the ordered and purchased deliveries between manufacturer and supplier, as well as the conditions of the general agreement, are known, the manufacturer can agree the payment and send advice of this to the supplier. He must then, of course, undertake an "invoice check" to ensure that the manufacturer's payment is consistent with the goods supplied. Hence, invoice checking is shifted from the manufacturer to the supplier. This sector has already been realized in the German automobile industry, and is therefore represented separately in Fig. C.III.21, where the bank function is also shown.

In general, the increased interdependence resulting from the improved information technology gives rise to a tendency for functions which where previously carried out by the supplier in the order processing context now to be taken over by the customer in the purchase planning context. This applies both between retail customers and manufacturers and between manufacturers and suppliers. In contrast, functions in the receipt of goods context are taken over by the supplier (for example, quality checks by the customer on receipt of goods are eliminated, to compensate the final inspection carried out by the supplier must be carried out more strictly). These tendencies are increased if the technological links are intensified as a result of inter-application communication.

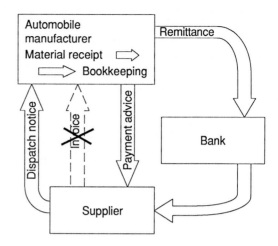

Fig. C.III.21: Transfer of invoice checking from manufacturer to supplier
Source: *Ford*

1.3 Application Integration

Fig. C.III.22 assumes that application programs can communicate directly with each other between the partners. This allows further reallocation of functions between the partners and a diminution in the amount of data exchanged.

The substitution between **ordering and order processing** is integrated, in that the customer (i.e. retail customer with respect to manufacturer, or manufacturer with respect to supplier) can also independently carry out the availability checks and hence material and deadline planning. Since the buyer enters the order into the system himself, he can also generate the confirmation according to his scheduling. Even the order handling right up to dispatch notice can be carried out by the purchaser, by accessing the production and dispatch planning data through direct enquiries. The invoice no longer needs to be sent to the customer, rather, the manufacturer can himself effect payment using a direct debiting service through electronic clearing houses linked to the bank sector. The invoice checking must, however, still be carried out by the retail customer. In the **logistic chain between manufacturer and supplier** the link between purchasing and order processing also becomes closer. Here too, the manufacturer can independently carry out availability checks and order pursuance via direct access to the supplier's planning programs.

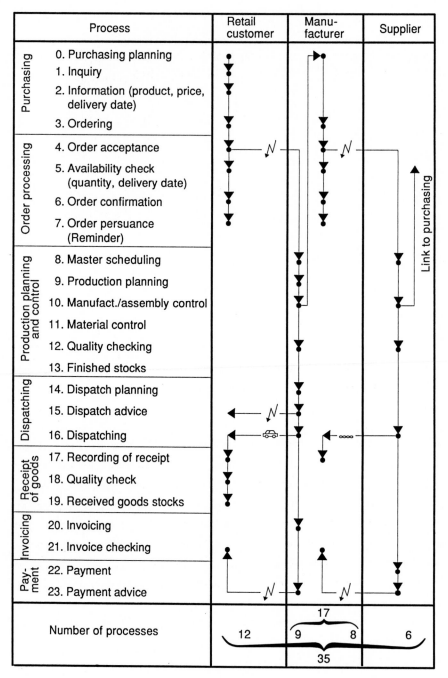

Process		Retail customer	Manufacturer	Supplier
Purchasing	0. Purchasing planning			
	1. Inquiry			
	2. Information (product, price, delivery date)			
	3. Ordering			
Order processing	4. Order acceptance			
	5. Availability check (quantity, delivery date)			
	6. Order confirmation			
	7. Order persuance (Reminder)			
Production planning and control	8. Master scheduling			
	9. Production planning			
	10. Manufact./assembly control			
	11. Material control			
	12. Quality checking			
	13. Finished stocks			
Dispatching	14. Dispatch planning			
	15. Dispatch advice			
	16. Dispatching			
Receipt of goods	17. Recording of receipt			
	18. Quality check			
	19. Received goods stocks			
Invoicing	20. Invoicing			
	21. Invoice checking			
Payment	22. Payment			
	23. Payment advice			
Number of processes		12	17 / 9 / 8 / 35	6

Fig. C.III.22: Data transfer reduction by shifting functions between partners

Depending on the strengths of the logistic partners the payment can be initiated either by the manufacturer or the supplier. For data transfer the manufacturer is considered to have the initiative. In application integration it is assumed that the supplier calls for payment once the delivery has been made. In this case "invoice checking" is transferred to the manufacturer.

1.4 General Effects

General effects of the increased information technological interdependence are:

1. Informational transfers between the partners in written form are drastically reduced.
2. The closer temporal integration between the operative functions manufacturing, assembly, ordering, delivery, etc, allows the elimination of planning functions between these sub-steps.
3. Functions are transferred between the partners. Typical examples are:
    -   Supplier's order processing functions are taken over by the customer,
    -   Functions relating to receipt of goods are transferred from customer to supplier.

These effects are evident from the example in the number of processes to be handled: **The number of processes is reduced from 50 in Fig. C.III.19 to 42, and then 35 at the highest level of informational integration**.

The substitution of functions must in general be achieved by greater provision of support from the unburdened partner. For example, the manufacturer must support the increased quality assurance functions of the supplier both procedurally and by transfer of know-how. In addition, he must support the closer temporal integration of the dispatching and purchasing systems through more detailed information about medium term production plans. This also means that master production plans must be agreed between manufacturer and supplier with a higher degree of validity. The appropriate technical prerequisites must be created for the information technical handling of planning functions. Where there is merely to be an exchange of documents (file transfer, electronic mail, etc.) the formats of the documents to be exchanged must be agreed. Fig. C.III.23 shows those documents for which the automobile industry has already established or recommended standards.

110

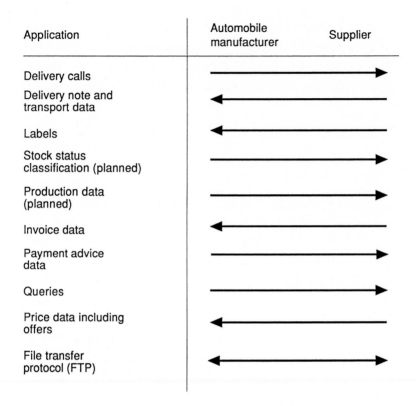

Application	Automobile manufacturer	Supplier

Fig. C.III.23: VDA recommendations for data transfer

The file transfer protocol is supported by a monitoring system, which is implemented on the automobile manufacturer's central computer and regulates the transfers. For data exchanges between the firms VW, Ford and Opel and their suppliers the monitoring system is RVS, for Daimler Benz it is the DAKS system. The VDA protocol recommendations are set out in the VDA recommendation 4914 (FTP = File Transfer Protocol).

Fig. C.III.24 shows a schematic representation of the transfer process (see *Schneider, Datenübertragung 1986*).

The manufacturer's central computer is generally defined as a so-called primary station, that of the supplier as a secondary station. Once the link with the primary station is established the secondary station must identify itself with an IDENT statement. The primary station then sends an IDENT statement back. Only when both statements have been accepted data is exchanged. Both partners can send data. The example only shows this for the primary station. The transfer begins with a header, in which the data records to be transferred are described, and a trailer which describes the data sent. The response

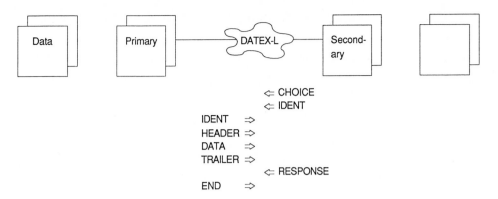

Fig. C.III.24: A transfer process

statement indicates transfer receipt. The files consist of records of a length of 50, 80, 129 and 152 bytes. Each record begins with the identification STX (start of transaction) and ends with ETX (end of transaction). The transfer is effected using the DATEX-L service with a BSC II protocol. In order for one supplier to be able to support the various postal services (DATEX-L, DATEX-P), producer networks (SNA, DECNET, ...) and producers' data display and formats (RVS, DAKS, ...) so-called remote transmission boxes have been

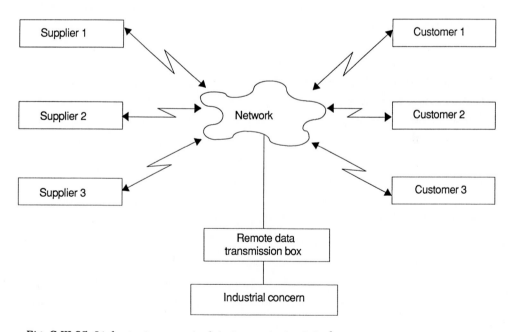

Fig. C.III.25: Linkage via a remote data transmission interface

developed, which create a uniform data display for the supplier by reformatting (e.g. the remote transmission box from the firm ACTIS). The perspectives shown in Fig. C.III.25, in which an industrial concern is linked via a remote data transmission interface with both customers and suppliers is no longer a utopian dream.

In the case of inter-application contact, computer programs must be able to communicate more intensively with each other. It is necessary, therefore, that the standardization of network services and operating systems must be extended to inter-application contacts.

2. CAD/CAM Integration

2.1 Initial Position

In the inter-company exchange of design data the relationship between the automobile industry and its suppliers (as well as the aircraft industry) is again most advanced (see *Schwindt, CAD-Austausch 1986*). In Fig. C.III.26 the traditional procedure is again shown first. Here it is assumed that both partners already have CAD systems in use, but that these are unconnected.

In the course of the rough design requirements for individual components are determined by the manufacturer. These are then specified in more detail in the preliminary design. The information from this preliminary design is transferred as a drawing to the supplier. He then enters these data anew and includes them in the detailed design, e.g. for tool construction, in accordance with his production-technical and technological possibilities. The result is then transferred back as a drawing to the manufacturer. The manufacturer can transfer the drawing to a stand-alone CAD system, and then test it in the course of design checking (e.g. calculations or simulations). Alterations are then once more transferred to the supplier as a drawing. On the basis of these alterations a final design is achieved, whose results are again made available to the manufacturer. A further control cycle can be added here, as is indicated in Fig. C.III.26.

Possible amendments to this initial position resulting from the two levels of information-technological integration "data transfer" and "inter-application contact" are examined below.

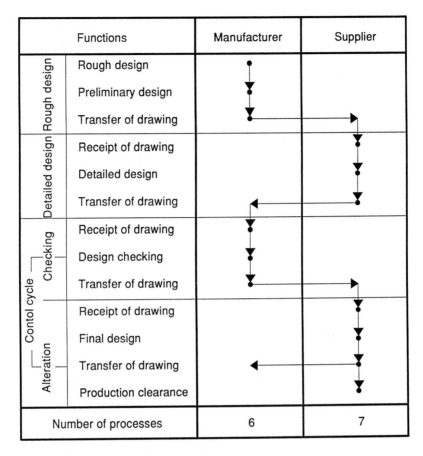

Functions		Manufacturer	Supplier
Rough design	Rough design		
	Preliminary design		
	Transfer of drawing		
Detailed design	Receipt of drawing		
	Detailed design		
	Transfer of drawing		
Checking	Receipt of drawing		
	Design checking		
	Transfer of drawing		
Alteration	Receipt of drawing		
	Final design		
	Transfer of drawing		
	Production clearance		
Number of processes		6	7

Fig. C.III.26: Utilization of unconnected CAD systems

2.2 Data Exchange

Fig. C.III.27 shows the process with electronic transfer of geometry data. This requires that CAD data can communicate via a common data interface. The IGES interface or the interface for free-form surfaces developed by the Verband der deutschen Automobilindustrie - VDA-FS should be cited here. The fundamental problem for such an interface is that the internal model representations of different CAD systems allow not only 1:1 relationships, but also 1:n or n:m relationships. Hence it is possible that not all information contained in one system can be transferred to the other system in the same form. For example, classification concepts cannot be transferred from one system to another. Assuming an appropriate interface, receipt of the geometry data no longer involves transfer of information from the written drawing documents into the CAD system. Hence the number of steps can be reduced from 6 to 5, and for the supplier from 7 to 5, disregarding the possibility of repeats of the control cycles between manufacturer and supplier.

	Functions	Manufacturer	Supplier
Rough design	Rough design		
	Preliminary design		
	Transfer of drawing		
Detailed design	Receipt of drawing		
	Detailed design		
	Transfer of drawing		
Checking	Receipt of drawing		
	Design checking		
	Transfer of drawing		
Alternation	Receipt of drawing		
	Final design		
	Transfer of drawing		
	Production clearance		
Number of processes		5	5

Fig. C.III.27: Process with electronic data transfer

2.3 Application Integration

Fig. C.III.28 assumes that the manufacturer can access data from the supplier in the inter-application communication context. This means that he can, for example, access the supplier's geometry data, bills of materials and technological data, and hence largely carry out the design himself.

Hence the control process is reversed: the design documents prepared by the manufacturer are communicated to the supplier and then checked by him as to their practicability. This procedure generally leads to the manufacturer taking over the supplier's design functions. This eliminates further data transmissions which were previously required. Fig. C.III.28 indicates that the number of processes for the manufacturer continues to be 5 (whereby new functional content is included, but data transfer functions are eliminated), whereas for the supplier they are reduced to three.

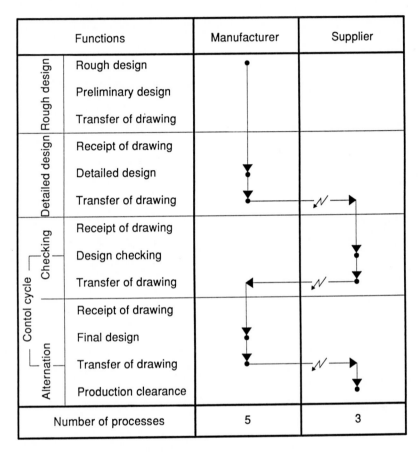

Fig. C.III.28: Access in the context of inter-application communication

2.4 General Effects

Strict pursuance of this tendency can lead to the entire design function of the supplier up to production clearance being taken over by the manufacturer. The manufacturer would then transfer the completed drawing as a data file, or even take over the NC programming and transfer complete NC data records. Using the same methods, he can also automatically obtain quality assurance data for control purposes from the measurement system of the CAM area. As a result of the close link between design and production

116

necessitated by the highly automated process, and high quality requirements necessitated by the temporal coupling of purchasing and production processes, the liaison can even go so far that the manufacturer not only carries out the design in accordance with the production-technological possibilities available to the supplier, but even exerts control himself over these technological possibilities. This can occur, for example, by prescribing or providing specific production equipment to be installed.

f. Sub-Chain 6: Linking Operative Systems with Accounting and Controlling Systems

The support for process chains within a CIM system relates primarily to the operative data processing level within the enterprise. However, the changes to the operative processes and their supporting data structures that CIM initiates also have their effects on the value-based accounting and controlling systems.

This is illustrated in Fig. C.III.29. The information pyramid shows how each operative system is assigned to a value-based accounting system within bookkeeping, on which further controlling systems up to enterprise planning are based.

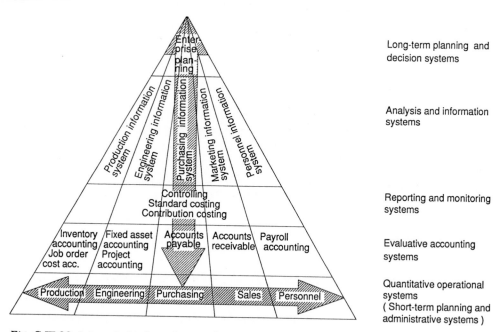

Fig. C.III.29: Integrated information systems

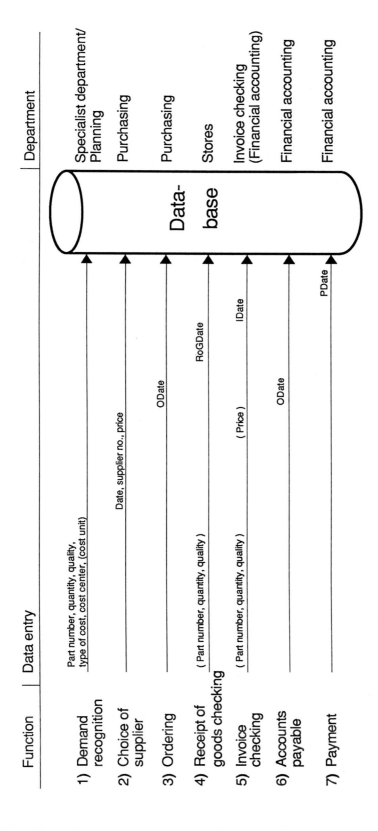

Fig. C.III.30: Data and process integration using the example of a purchasing process chain

Any business transaction, therefore, passes through both the quantitative and the valuational levels. This is illustrated by the example of the simple case of order handling shown in Fig. C.III.30.

The process chain begins with the specification of requirements and proceeds via the choice of suppliers to order writing. The goods received, accompanied by a delivery note, are subjected to a receipt of goods check. On receipt of invoice the invoice is checked against the order and the receipt of goods. After invoice checking the process is passed on to accounts payable, where the bookkeeping entry is made and payment of the invoice is arranged for.

This kind of process chain can extend over a considerable time period (often several months) and, as is shown on the right side of Fig. C.III.30, can be processed by several departments.

In spite of the long timescale, the involvement of diverse areas of organizational responsibility, and the inclusion of diverse documents (order form, delivery note, invoice, transfer slip) it is merely **one** process that is being considered.

The progressive accumulation of data is represented graphically in Fig. C.III.30. It is apparent that the later functions within the process chain merely make correction or confirmation entries (indicated in Fig. C.III.30 by the respective date entries).

As soon as the choice of supplier has been made the order price is established. Consequently, practically all the data needed for the production of the invoice and even the accounts payable entries are already available. Once the order has been finalized, therefore, a "pro forma" invoice, including a "pro forma" entry record can be generated. Accounts payable can take over these data, so that the operative order handling processes and the accompanying accounting procedures are integrated. At the same time, these data form the basis for stock evaluations in the inventory control context.

In addition to the designing of CIM process chains at the operative level, which necessitate integration between functional areas (see the horizontal arrow in Fig. C.III.29), the further utilization of data in the information systems superimposed on the operative systems is therefore also a CIM component.

The evaluative accounting systems superimposed on the quantitative systems should have direct access to the data from the operative systems. Furthermore, the quantitative systems should also keep a record of those values that have already arisen in connection with the operative processes. The same requirements apply to the consolidation into control systems on up to planning and decision support at the enterprise level. This integration principle is indicated by the vertical arrow in Fig. C.III.29.

The interface between the quantitative, operative systems and the evaluative, accounting systems is the most obvious, since here each functional planning system can be assigned an evaluative counterpart. The principle of data processing integration has in recent years led to close interdependence of the information systems in this context. This means, for example, that accounts receivable and payable are directly provided with entry records from the sales and purchasing systems. Amendments within the sales and purchasing systems, which are effected by inter-company data exchange within the CIM sub-chain, therefore, take effect directly in the corresponding accounting system. A possible implication of this is that in close temporal call-off systems between customer and supplier traditional documents, such as invoices and delivery notes, can be eliminated completely. Since prices have already been established in the context of framework agreements, and call-offs are arranged at short notice by the producer (customer) with the supplier, an invoice which is then sent by the supplier to the customer has no informational content, and can, therefore, in principle be dispensed with.

The close data interdependences between the operative and the evaluative levels are not only apparent within the sales and purchasing systems, but also between pay-roll accounting and operational data collection, for example. The time and performance data needed for gross wage calculations are increasingly obtained from the operational data collection system. This means that operational data collection systems should not be designed simply on the basis of a technically construed CIM model, but also bearing the evaluative accounting levels in mind. Cost accounting is also an important recipient of information for parallel order accounting. The nature of the problem is made clear in the example of a CIM architecture in which a decentralized production control system is designed for an autonomous sub-area (e.g. a production island or flexible production system) in which only key values (deadlines, quantities, qualities) are passed on from the higher level order creation system to the independent unit, which then carries out its own differentiated planning functions on the basis of these key values and generates the corresponding data.

However, if alongside this decentralization a centrally organized system for pay-roll recording and cost accounting is maintained which expects a different kind of feedback message from that envisaged by the decentralized control system, discrepancies arise within the system design: PPS is decentralized, cost accounting and pay-roll accounting continue to be centralized.

These examples should make it clear that only collective structuring of the quantitative and evaluative systems can ensure that the data display the required consistency, freedom from redundancy, and currency.

CIM's concentration on the operative processes has obscured the view of the evaluation and analysis systems. Databases have been constructed "from ahead" as it were, so that they can function as planning data to control the operative processes. To make these data also available for analysis and evaluation purposes, remains a largely unexplored undertaking. This applies not only to quality assurance systems, but also to the establishing of standard times, quantities and costs so that appropriate planned/actual analyses can reveal shortcomings in the planning procedure or in the data. Expert systems can also be useful here, in that they can pursue intelligent checking procedures for planned/actual comparisons on certain data constellations for diverse process chains and databases (see *Scheer, Kraemer, Konzeption und Realisierung II 1989*).

The enterprise planning level is also influenced by CIM. In general, it tends to be the case that while the design of CIM systems is of considerable complexity the resulting operation is correspondingly simplified as a result of the high degree of automation. This means that the preparation of information for the structuring of the system will be of increasing importance. Simulation models can be employed to ensure the optimal structure of CIM systems, which indicate, for example, the effects of a flexible production system incorporating automated material flows on inventory levels and equipment location requirements.

IV. Critical Success Factors

The significance of the individual CIM sub-chains within a CIM concept depends on the production structure of the industrial concern in question. A general weighting is shown in Fig. C.IV.01.

With regard to the production structure a simplifying distinction is made between unit production and process-oriented manufacturing. The term unit production refers to the structure of a manufacturing concern in which complicated units are produced, often with the help of deeply layered bills of materials. Process-oriented manufacturing, on the other hand, refers to those structures which are typical of the chemical, food and paper industries. Here few inputs are used to produce a multiplicity of finished products from one manufacturing process. Often the unit of measurement chosen will be the weight of the output.

Within these two groups further differentiation can be made, for unit production between one-off production, order and small series production, and large series and mass production, and for process-oriented manufacturing between order and small series and large series and mass production. One-off production in process-oriented manufacturing is not considered, since this is atypical of this manufacturing form.

Legend:
- ◯ = High
- ○ = Medium
- ● = Slight
- | = Zero

	"Parts production"			Process oriented production	
	One-off production	Ordered small series production	Large series/mass production	Small series orders	Large series/mass production
1. sub-chain: Planning-control	High	High	Medium	High	Medium
2. sub-chain: CAD/CAM	Medium	High	Slight	Zero	Zero
3. sub-chain: Primary data management/ product development	High	High	High	Medium	Medium
4. sub-chain: CAM-production control	High	High	Medium	High	Medium
5. sub-chain: Inter-company data exchange	Medium	Medium	High	High	High
6. sub-chain: Operative level-evaluative level	High	High	High	High	High

Fig. C.IV.01: Weighting of CIM sub-chains

The first sub-chain describes the **specification of a CIM hierarchy**. This must be regarded as fundamentally important since the basic planning and control functions are here distributed to a computer architecture. The importance of establishing a strong hierarchy increases if the decentralized units are given considerable freedom in optimizing the manufacturing process, such that independent scheduling functions can occur within order supply from a higher hierarchy level. With large series and mass production the manufacturing process is already so strongly optimized (e.g. in the use of production lines with strict timing) that local scheduling largely disappears. For this reason the process of establishing the planning and control hierarchy in this case receives middle weighting.

The **CAD/CAM link** is especially important in unit production structures with flexible order and small series manufacturing. With large series and mass production forms, however, the relationship is not so important, since here design adjustments can be planned more in advance and are less dependent on a strict data flow. In process-oriented manufacturing this chain is disregarded, since CAD systems do not have the same importance here (e.g. the chemical industry) as in unit production.

The **primary data management chain** for bills of materials and work schedule information and the support of product development is of considerable importance throughout unit production. In process-oriented manufacturing this applies to control instructions which are administered both by development and production control, and to the associated work schedules. Nevertheless, it is of somewhat less significance than in unit production.

The links from **operational data collection** and **process control** to the control of computerized manufacturing installations are of primary importance in large series and mass production within unit production, as well as in the entire range of process-oriented manufacturing. In one-off production procedures, order and small series manufacturing, on the other hand, automatic data transfer, e.g. for order related information, is not easily achievable, since the automatic transfer of order related information (e.g. quantity and quality) depends on a high degree of standardization of the manufacturing process.

Inter-company data exchange is at present primarily of significance for series production in the automobile industry. However, this experience will then be transferred to other industrial sectors.

The **linking of operative and evaluative information systems** is of considerable significance in all sectors.

The weighting shown in Fig. C.IV.01 is not exhaustive. It is merely meant to indicate that differences in the central CIM issues can arise as a result of differences in the manufacturing structure of the enterprise.

A further influential factor in the weighting of process chains is the fundamental strategic approach of the enterprise.

Approaches to a systematic investigation of the strategic importance of information technology have been discussed above all in the USA. In his approach, Porter has investigated how the situation within the enterprise is affected by

- suppliers,
- customers,
- new competitors,
- product or service substitutes,
- market rivalry

(see Porter, *Creating and Sustaining Superior Performance 1985*). This analysis uncovers the strengths and weaknesses of the enterprise, and can be used as the basis for further investigation.

This kind of analysis shows that general strategic aims, such as

- cost leadership,
- flexibility with regard to customer wishes and
- exploitation of market position,

make differing demands on the CIM concept, and hence place different CIM sub-chains in the foreground. An enterprise which puts particular emphasis on cost leadership will pay more attention to the CAM system, i.e. process automation, while an enterprise which aims primarily at market related flexibility will attach more importance to the CAD/CAM chain. The latter would then be in a position to pass on customer demand-oriented variants in the shortest time from the order acceptance system through design to production.

Rockart's concept of "Critical Success Factors" (CSF) describes another approach (see *Rockart, Chief Executives Define their Own Data Needs 1979*). A CSF defines an area or an activity which **must** be improved. On the basis of this definition strategies can be developed which fulfil the enterprise's goals. For manufacturing industry Rockart names the following as CSFs:

- production planning,
- penetration of foreign markets and
- production automation.

124

The concept of critical success factors can also be related to the CIM problem area from other standpoints. Assessing the economic efficiency of introducing a CIM sub-chain is extremely difficult. The costs of the computer system and its organizational introduction can be estimated with the most accuracy, even though considerable uncertainty results from sudden price rises and inaccuracies regarding installation dates.

Calculating the benefits is much more difficult. Firstly a multiplicity of so-called qualitative influencing variables must be taken into account. A further problem arises from the fact that benefits often accrue at a different point from the investment costs. For example, the investment costs of introducing a CAD system are incurred by the design department (see Fig. C.IV.02).

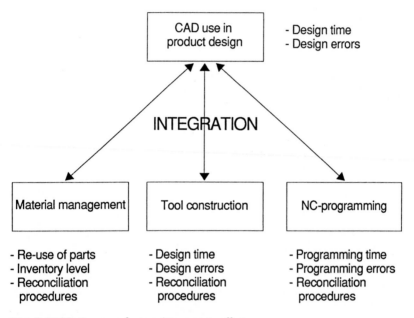

Fig. C.IV.02: Success factors/Economic efficiency

If the economic efficiency is only assessed on the basis of whether the costs are more than offset by the resulting reduction in the need for designers the result will generally be negative.

Only when the advantages to other organizational areas are taken into account will the introduction of the system appear profitable. For example, the automatic transfer of product geometry to toolmaking can simplify considerably the design of machine tools. The transfer of bills of materials information from CAD supports process planning. The better use of classification systems and the resulting improvement in the search for similar parts can facilitate a reduction in the parts spectrum. This in turn reduces the planning effort required in the material management context, and reduces stock levels. The transfer of the product geometry to NC programming provides further support for process planning.

These examples show that only complete consideration of the entire "geometry flow" process chain can allow accurate assessment of economic efficiency. In order to be able

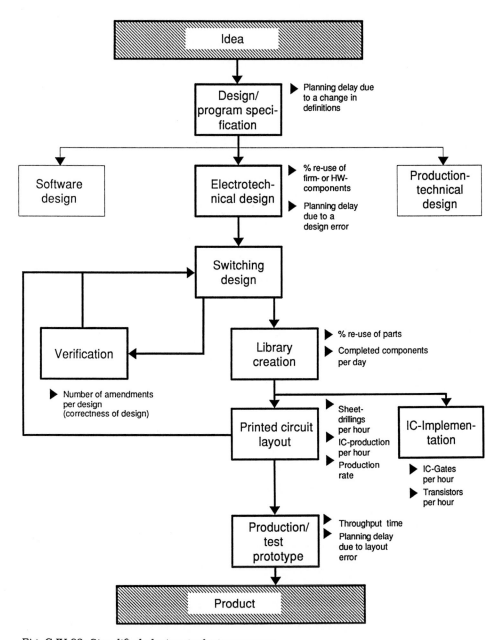

Fig. C.IV.03: Simplified electronic design process
from: *HP*

to really exploit such effects success factors need to be defined for the introduction of this kind of system, which first make clear the objectives of the introduction, and which are then available to management as planned values for monitoring the success of the introduction. Fig. C.IV.03 illustrates this in detail for the introduction of a CAD system in electronics manufacturing.

Fig. C.IV.04 generalizes this approach in that goals are first more clearly specified by means of an interpretation, and then rendered operational by success factors for goal targets and their monitoring.

Goals	Interpretation	Success factors
Productivity	General improvement in manufacturing productivity	Internal rate of interest Turnover per head
Quality	Improved total quality costs Reduced product complexity	Guarantee costs as a proportion of turnover
Flexibility	Improved ability to react to market changes	Reduction of lot size Set-up time Order throughput time
Pace of development	Exploitation of market opportunities Improved sensitivity to customer wishes	Market share Time for prototype creation Development time for a new procedure Punctual delivery
Costs	Reduction of unit total costs	Manufacturing costs per unit

Fig. C.IV.04: Success factors for CIM

V. CIM Function Levels

An essential component of the CIM concept is the creation of the organizational pre-requisites. This primarily means specifying the CIM functions within an organizational hierarchy.

In many enterprises in the CIM areas of production planning, production control, design and manufacturing a multitude of hardware and software systems are currently springing up. Because in large concerns these areas are often organizationally autonomous, a variety of solutions is often developed even within the individual factory. This can lead to the situation in which the same function, e.g. the back-up of the operational data collection system, may be carried out in one factory on a dedicated local computer, and in another factory on the host general purpose computer. This uncoordinated functional break-down between different levels of a computer network, also associated with different computer environments (local systems, database systems, programming languages, etc.), cannot make sense within a closed CIM concept. For this reason the following discussion will develop a model CIM functional hierarchy for an industrial concern. Finally, the allocation of basic tasks to specific levels of the hierarchy will be elaborated. Here it will be shown that these tasks are similar for each level, so that typical function modules can be developed.

a. Levels Concept

Given the comprehensive character of CIM the following discussion is relevant to all the hierarchical levels of an industrial concern. These hierarchical levels include, in the production context:
- company headquarters,
- product group level,
- factory level,
- factory area level,
- equipment group level,
- machine level,
- machine component level.

In the sales market context the hierarchy levels might be:
- branch office,
- selling agency,
- sales.

It is typical of the hierarchical structure that each entity of one level comprises several entities of the directly subordinate level. A factory can be used for several different product groups, so here several arrows may join together. A branch office can also be responsible for several product groups. All arrows indicate incoming or outgoing information paths. Fig. C.V.01 depicts examples of important computerized functions in the CIM areas of sales, PPC, design and manufacturing. It must be emphasized that these functional assignments do not automatically imply a corresponding computer hierarchy. The structuring of the computer levels will be discussed below, and is based primarily on computer-technical criteria such as system availability, response time characteristics, data transfer times (and costs) or possibility of using peripherals.

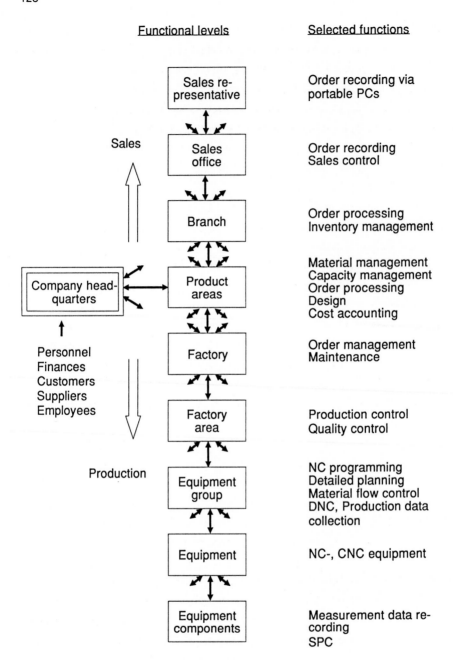

Fig. C.V.01: Functional levels of the logistic chain

Which functions a level is responsible for must be established first, however, since this determines the type and extent of computer support required.

The function assignment should also ensure that an enterprise is aware of the requirements concerning quality, currency, exploitation of synergy effects and the room for manoeuvre within organizational lines, so that they can be taken into account in the structuring of the information system.

For example, if operational data collection is assigned to the factory level this leads to a (factory-)central collection office, to which all the production points deliver their records (pay slips, order papers). This inevitably generates redundancies and delays. In contrast, if the same task is assigned to the equipment group a (networked) terminal system can ensure much greater temporal currency.

In the functional breakdown it is also necessary to differentiate between the responsibility for structuring and responsibility for executing the function.

The architecture presented in Fig. C.V.01 will be explained briefly.

Strategic planning functions for establishing product and production strategies are assigned to the **company headquarters**. At the same time global business functions, such as payment settlements, wage and salary calculations and controlling are also located here. It can also make sense to handle purchasing at the central level because of its global function. Essential primary data relate to personnel, creditors and debtors. At the same time, product and material data are kept centrally.

Essential CIM output data, which are passed on to the next hierarchy level are for production the strategic product plan and the production plan, and for marketing the strategic sales plan.

At the **product group** level the primary data for bills of materials, work schedules and equipment are managed. The technically-oriented function assigned to this level is design (CAD). The more PPC-oriented functions at this level are sales, master planning, material management and capacity management. **Output from this level are the released sales and production orders and released drawings (designs).**

The **branch offices** process the orders recorded by the selling agencies and manage stocks on hand (at the branch office). The selling agencies control field service and take over the orders acquired by the sales representatives. In order to achieve more up-to-date order processing, function levels can also be technically by-passed in the course of order processing.

At the **factory level** the released orders from material and capacity management are handled. Typical primary data lists handled at the factory level are tool and device data and employee data. More technically-oriented computer functions which can be allocated to this level are equipment design and maintenance control. Storage systems may also be set up. In the production planning and control area the detailed scheduling of orders received and the associated task of allocation to various factory sub-areas is carried out. **Output from the factory level are released production orders at the operation level.**

Production and quality control functions are assigned to the **factory area** level. **Output from the factory area level are conveyance and production operations and stock movements.**

Various **manufacturing areas** can be defined within the factory area. On the basis of recent trends towards the decentralization of production, a growth in self-regulating sub-areas should be reckoned with. These may be flexible production systems, production islands, Kanban routes, processing centers, assembly islands, etc. Here too, the primary task is to manage orders received. Primary data for special tools may also be managed here. More technical functions are NC programming, DNC operation and the control of area-specific conveyance and storage systems. In the PPC area cutting and optimization problems, most detailed scheduling of operations taking combining and splitting into account, sequencing and equipment allocation are important. At the same time operational data collection functions can be carried out. **Output from this hierarchy level are released orders to specific equipment specifying operations, concrete conveyance orders, released NC programs, etc.**

At the **machine** level the operations to be carried out are managed. Technical functions are NC and CNC operations to be executed. Operational data can be directly obtained from the controls by the scanning of weighing and counting procedures. **Output from this hierarchy level are concrete control instructions to machine components.**

At the **machine component** level dedicated controls are employed, which can, for example, undertake constant collision checks on the driverless transport system.

The functional allocation to enterprise hierarchies given here is only an example. It has been chosen, though, to cover the widest possible range of applications, and to draw attention to the most important arguments with respect to computer and data accessibility. It can, therefore, serve as a starting point for internal discussion of such allocation issues.

b. Tasks of a CIM Level

Within the pre-defined CIM hierarchy, basic computer-technical and operational structuring functions are carried out at each level, connecting it with the next level up. Since, at present, the hierarchy shown in Fig. C.V.01 has scarcely been implemented, even these basic functions have not yet been clearly worked through. Rather, it is typical of present industrial enterprises that, although a certain orderly division between company headquarters and product areas exists, within the various factory levels there scarcely exist any discernible orderly regulations. For this reason, the question of a standardized modularization of functions has up to now received little discussion. Fig. C.V.02 shows such a developed modularization. Its suitability can be assessed using the functions specified at each level in Fig. C.V.01.

First of all, each level must carry out data management functions for their own primary data and for data obtained from upper and lower levels. The essential functional tasks consist of transforming input data from higher levels using planning and control functions and passing them on to lower levels, and taking feedback messages from the lower levels, transforming them, and sending them on to higher levels. These two transformation paths are shown separately in Fig. C.V.02.

In the direction of the information flow from above to below, a check on the priority of acquired input data is first carried out. Here it is necessary to distinguish between standardized (periodic) data transfer and event-related data transfer (e.g. from a rush order). The transferred values are entered as input information in the level-specific planning and control functions. Finally, data to be supplied to the lower levels are determined. Here data may be produced which by-pass several hierarchy levels. To safeguard a controlled information flow, however, it is worthwhile specifying clear and simple hierarchy routes for such data, too.

Before order data (in the widest sense) are passed on to a lower level, a check is made as to whether the resources to be used are at hand at the lower level (availability check). These checks need only be carried out at that level of detail which is adequate for passing on the information. For a concrete implementation recommendation at the next level, a more detailed availability check can be carried out. After successful availability checks on resources (e.g. equipment, work force, NC programs, materials) key data for the subordinate level are defined. Such key data might consist of quantities, times, qualities, tolerances, etc. within which the subordinate level must operate.

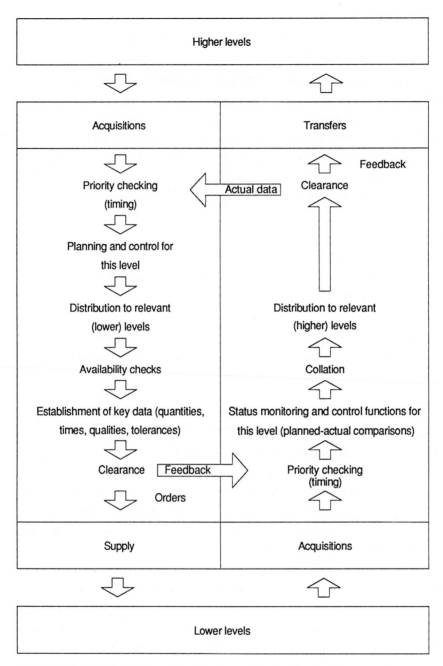

Fig. C.V.02: Modularization of the functions of a CIM hierarchy

It is part of the basic philosophy of the hierarchical model that the lower level is always allowed as much self-determination of control functions as is consistent with the maintenance of the key data received from the higher level. In this sense the lower level is regarded by a higher level function in the role of "external suppliers".

With the release of this type of output information the lower levels are provided with the necessary input information. At the same time information may be produced which constitutes receipt or feedback data for higher levels. These data are forwarded as "feedback" in the second information flow, which runs from the bottom up to the top. Here, data from lower levels are first received and checked against priorities. This check activates the necessary temporal control of the received data. Receipt of a disruption message which necessitates immediate intervention of the higher level control mechanism will, for example, initiate a real-time action, while routine receipt of feedback information (e.g. completed jobs from the data collection system) requires no real-time processing. The data received are subjected to level-specific status control and planned/actual comparisons. Before being passed on to higher levels the data are condensed and prepared for various higher level processes. For example, information from the operational data collection system can be used at the factory level for cost accounting functions at that level, for wage calculation functions at the factory and company headquarters level and for representing the actual situation to form the basis of the production control system at the factory area level.

The information obtained is released as feedback to higher levels. Actual data relevant to the same level are passed on to the control function of the information flow on the left. In this way the two perpendicular information flows form an information circulation system within the level. In total, the connections between levels form a system of cascading circular flows.

In currently available CIM components isolated functional processing of the individual level is dominant. The receipt and supply functions of prior and subsequent levels receive, in contrast, little emphasis (consider, for example, the forwarding of geometry data to the bill of materials management system and NC programming, or the transfer of data between production **planning** and production **control** functions or the independent control of a flexible production system). It is to be expected, however, that further developments of CIM systems will accord more weight to these coordination functions. Within a PPC architecture, for example, the emphasis will tend to shift from the more typical planning-oriented functions of material and capacity management to short term control functions. Within the more technically-oriented CAD/CAM functions a greater integration between geometry and technology-oriented processing will occur.

VI. Data Structures

An integrated database is necessary to support the process chains. The design of the logical structure of this database is therefore one of the most important steps in the construction of a CIM system. Within the logical data structure the necessary data objects (entity types) and their relationships to each other are designed. This step forms the interface between the specialist knowledge of the CIM user and the formalization requirements of the information technology. Errors made during this design stage, can scarcely be corrected in the following implementation stages.

Chen's Entity Relationship Model (ERM) provides an effective design language for the creation of logical data structures. It has already been employed to represent the primary data management of PPC and for the product development chain.

The elements of the ERM are entity and relationship types (see Fig. C.VI.01). An entity type or object type is described in the database context by attributes. Typical examples of entity types are orders, equipment, tools and employees. Relationship types represent the links between the entity types, for example, the assignment of orders to equipment or the assignment of employees to orders. The relationship types can be differentiated according to the degree of interdependence into 1:1, 1:n and n:m relationships.

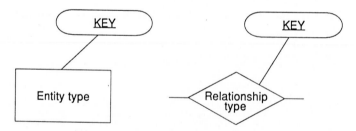

Fig. C.VI.01: Entity and relationship types

The informative value of the logical data structures can be shown using two examples for product description and production control.

The use of the Entity Relationship Diagram to support the "product description" process chain has already been shown in Fig. C.III.12. This is represented in more detail in Fig. C.VI.02.

The bill of materials structure is represented using the entity type PART and the relationship type STRUCTURE. Geometric forms are described using the entity type FORM. The relationship "belongs to" (PNO, FNO) assigns to a part several geometric elements which characterize it. In the same way one geometric form can enter into

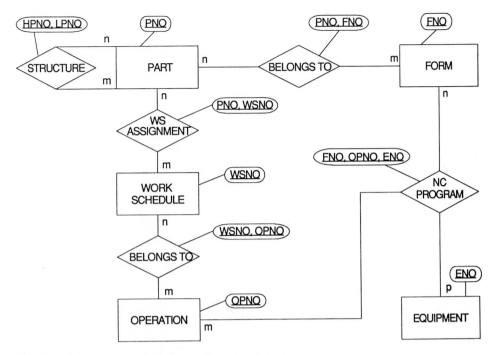

Fig. C.VI.02: Data structure for product description

various parts. The relationship type therefore constitutes the link between the geometry data of a CAD system and the bill of materials management of a PPC system.

The production prescriptions are described in the WORK SCHEDULE entity type and the associated OPERATION entity type. The n:m relationship "WS-ASSIGNMENT" between PART and WORK SCHEDULE indicates that a part can be produced using various work schedules (e.g. depending on the quantity to be produced) and that one work schedule can be used to produce various parts (where the differences consist merely of slight bill of material variations). The n:m relationship between OPERATION and WORK SCHEDULE also contains great data flexibility, which simultaneously allows low-redundancy data management.

In automated production systems which are driven by control programs the control instructions relate to the part to be produced (characterized by the geometry), to the machine and to the operation to be executed. For this reason the NC program is defined as the relationship type between FORM, OPERATION and EQUIPMENT (machine).

The data structure of Fig. C.VI.02 thereby allows an integrated data flow between material management, design and work scheduling including NC programming. Since the sub-areas in question are at present (still) in general supported by independently developed computer systems, in any concrete CIM implementation the main concern is to

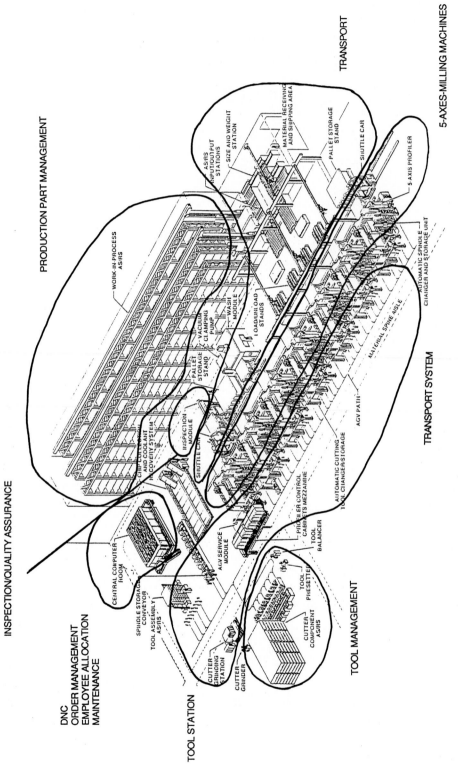

INSPECTION/QUALITY ASSURANCE

PRODUCTION PART MANAGEMENT

TRANSPORT

5-AXES-MILLING MACHINES

DNC
ORDER MANAGEMENT
EMPLOYEE ALLOCATION
MAINTENANCE

TOOL STATION

TOOL MANAGEMENT

TRANSPORT SYSTEM

WORK-IN-PROCESS AS/RS

AS/RS INPUT/OUTPUT STATIONS

SIZE AND WEIGHT STATION

MATERIAL RECEIVING AND SHIPPING AREA

PALLET STORAGE STAND

SHUTTLE CAR

5-AXIS PROFILER

AUTOMATIC SPINDLE CHANGER AND STORAGE UNIT

MATERIAL SPINE AISLE

AGV PATH

AUTOMATIC CUTTING TOOL CHANGER/STORAGE

PRIMARY CONTROL CABINETS MEZZANINE

TOOL BALANCER

TOOL PRESETTER

CUTTER COMPONENT AS/RS

CUTTER GRINDER

CUTTER GRINDING STATION

TOOL ASSEMBLY AS/RS

SPINDLE STORAGE CONVEYOR

CENTRAL COMPUTER ROOM

AGV SERVICE MODULE

WASH MODULE

SHUTTLE CAR

CHIP, COOLANT AND COOLANT RECOVERY SYSTEM

INSPECTION MODULE

PALLET STORAGE STAND

VACUUM CLAMPING PUMP

WASH MODULE

LOAD/UNLOAD STANDS

Fig. C.VI.03: Flexible production system

create the linkages between the data structures. The prior logical penetration of the data relationships is also an essential prerequisite for this task.

In the support of the CIM sub-chain "decentralized production control" diverse application areas interact with each other. Fig. C.VI.03 represents a flexible production system (FPS).

Quality assurance, production control, NC programming, tool management, material management, tool conveyance, material conveyance and maintenance are integrated into a single system.

Before an operation on a 5-axle milling machine can be started

- the operation must be released by order control,
- the NC program must be available,
- the appropriate tool must be taken from the tool inventory and conveyed to the machine,
- the production part must be taken from the parts inventory and conveyed to the machine,
- the test plan for quality assurance must be made available.

Only a logically integrated data structure ensures that these data are simultaneously available (see Fig. C.VI.04).

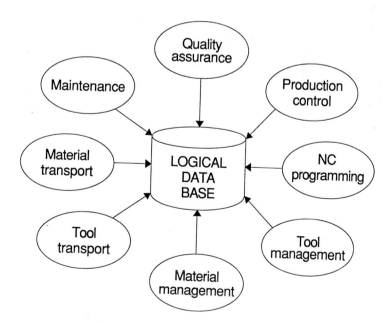

Fig. C.VI.04: Integration via a logical database

At the heart of this data structure is the MACHINE LOADING, which consists of the elements OPERATION, EQUIPMENT, SEQUENCE and TIME (see Fig. C.VI.05).

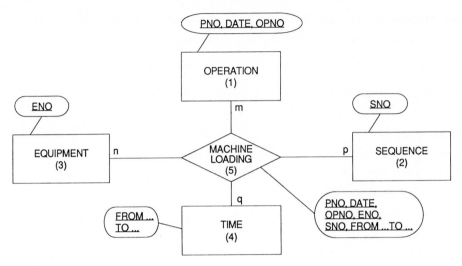

Fig. C.VI.05: ERM for detailed scheduling

The entity type SEQUENCE describes a collection of diverse (sub-) operations, which constitute a planning unit, e.g. as a result of material cutting optimization or order combination. The entity type TIME specifies the scheduling information by giving start and end deadlines.

An impression of the complexity of the data structure in production is given in Fig. C.VI.06, which presents the data structures not only for product description but also for the other production-related areas shown in Fig. C.VI.03 (see *Scheer, Enterprise-Wide Data Modelling 1989, p. 315*).

The logical data structure constitutes an integration framework into which the individual sub-systems can be incorporated. Here the interface problems become apparent, since in general inventory, conveyance and tool management systems are obtained from different suppliers.

The ERM diagram of Fig. C.VI.06 shows that with the help of this descriptive language very complex interdependences can be represented in a comprehensible form. The rigourous development of the CIM data structure generates an enterprise data model. The model developed by Scheer (see *Scheer, Enterprise-Wide Data Modelling 1989*) consists of about 300 entity and relationship types and depicts the interrelationships between production, engineering, purchasing, sales, personnel, financial accounting and controlling. It has already been used several times as a reference model for the specification of individual enterprise-specific models.

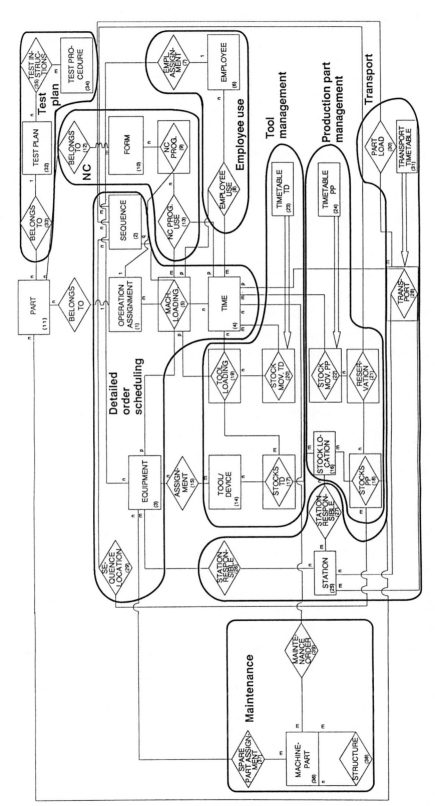

Fig. C.VI.06: ERM for integrated CAM systems

VII. Applications Software

The implementation of integrated process chains assumes appropriate applications software. It should first be established whether suitable standard applications software is available. Where in-house developments are required pre-fabricated software components should be employed. For the support of integrated processes at the workplace where the processing of tasks is subject to greater demands, the use of expert systems in the CIM context is of particular significance.

a. Standard Software

Requirement profiles are developed for the applications software to be used to handle the process chains that have been generated. Given the well-known fact that software development costs have risen drastically, the possibilities for using suitable standard software should be investigated very thoroughly. In order to support integration principles, the use of standard software families (altered as little as possible) is to be recommended so that the data and functional linkages incorporated by the software developers can be fully exploited.

Fig. C.VII.01 shows an excerpt from a requirement profile for the choice of a production control system. In evaluating concrete standard software products using check lists it is possible to distinguish between "must" properties which have to be fulfilled in all cases, and "should" and "can" properties which, although desirable, are not absolutely essential to the overall success of the system.

Integrated standard software is available above all for the upper part of the left side of the Y-diagram, that is for the more operationally-oriented planning functions. Here standard software families are available which display considerable integration with the operational accounting and controlling functions (e.g. the product family R from the software house SAP in Walldorf, the software from the house of Steeb for medium sized EDP systems or the IBM software MAPICS).

It is true that there exists a relatively large choice of standard applications software for the other CIM components, but these are merely available for sub-components, being offered by a multiplicity of diverse, sometimes smaller, software houses (see *Hellwig, Hellwig, CIM-Konzepte 1986*). Their integration, therefore, raises considerable problems. Integration possibilities for heterogeneous applications software and development trends will, therefore, be considered in more detail below.

	"Must"	"Should"	"Can"
Determinig the planned duration of an order on the basis of:			
- Set-up time	yes		
- Piece rate	yes		
- Dismantling time	yes		
- Quantity processed			
- Rejects			
- Order quantity	yes		
- Performance index: setting-up		yes	
- Performance index: processing			no
- Procedure			
Determining the planned appointed time for an operation on the basis of:			
- Loading duration	yes		
- Laytime (process related)	yes		
- Minimum forwarding quantity	yes		
- Earliest starting date	yes		
Latest starting date		no	
st finishing date	yes		
is	yes		
	yes		
			no

Fig. C.VII.01: Check-list for applications software for production control

b. In-House Development Using Pre-Fabricated Components and Standards

The use of standard software is problematic in the case where distinctive individual applications concepts have been developed in the firm. This might apply, for instance, to sales-related functions if special service functions have been developed for customers (conditions, possibility of choice between variants). If special organizational or production-technical knowledge exists in the production area, then individual software developments may be necessary here, too.

If, on the basis of these factors and after exhaustive checking of the possibility of using standard software, in-house development is necessary even here pre-fabricated components should be used wherever possible in order to keep development costs as low as possible. Similarly, advanced software development tools (CASE - Computer Aided Software Engineering) should also be used.

An interesting model for the creation of production related systems is currently being developed by IBM within the "Enabler" concept (see Fig. C.VII.02). Standard functions for the creation of applications software in the production control area are provided in the form of a toolbox, from which the user can put together his individual applications program. Similarly, communication elements are available for linking control functions at the machine level.

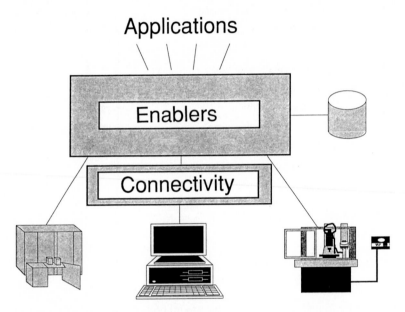

Fig. C.VII.02: Enabler software principle
Source: *IBM*

In addition to these and similar producer-specific concepts open standards can also be used to link applications software components. Fig. C.VII.03 shows the embedding of the standard software system FI-2 for production control at the heart of such a conception. Standard functions for order management and detailed control are carried out by the control center. This is extended by software components, which generate the communication with diverse higher level PPC systems regarding order supply, and also by communication components providing links to the CAM components.

Since it is clear that standard software in the CAM area (e.g. for operational data collection, quality assurance, etc.) is also increasingly using open standards (UNIX, SQL,

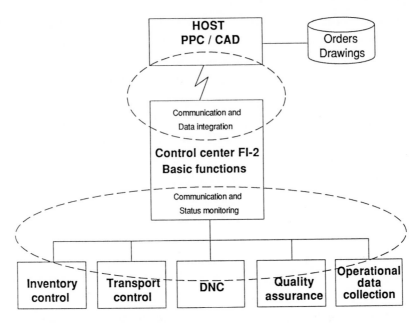

Fig. C.VII.03: Integration via open standards

the programming language C and X-windows) easier linking of independently produced applications software is possible.

Diverse applications systems can be linked using the concept of a CIM data handler, which is described below.

c. Expert Systems in CIM

1. Structure and Areas of Application

(Dipl.-Wirtsch.-Ing. Dieter Steinmann, Institut für Wirtschaftsinformatik (IWi), University of Saarbrücken)

Expert systems should support the decision maker in handling poorly structured decision problems (see *Scheer, Entscheidungsunterstützungssysteme 1986*). Poorly structured means that the determining factors for the decision are unknown, very large in number or subject to impenetrable interdependences. Expert systems can provide support in two situations: first in the choice of relevant and available information, second, in the choice of tools which can be used to process this information, and in the interpretation and evaluation of results. Information may consist of data or models representing a section of some real operation.

Given the complex structures resulting from the availability of information from diverse functional areas and the aim of handling process chains that are as self-contained as possible, there are also areas of action and decision within CIM which are difficult to structure. Examples of poorly structured decisions in CIM arise in the following functional areas: in preliminary pricing, for example, the configuration of complex products or customer orders, the planning functions within design, and the tasks of the purchasing areas. This is although the case in the creation of bills of materials and work schedules, since there accrues here a multiplicity of information about production procedures (depending on materials and required goods), available capacity (alternative work schedules), individual job operations, quality and cost information, etc. In the planning functions area, despite the support of conventional software products, the result depends very largely on the abilities of the employee responsible. The number of situation-specific alternative solutions and the consequent solution paths and alternatives are often so great that conventional systems with their systematic evaluation of all solutions are overstretched.

Up to now in decision processes for work schedule creation systems based on decision tables have mostly been employed. Their approach is similar to that of expert systems, but they do not possess the same flexibility or efficiency. The systems that have so far been available for generating work schedules or bills of materials are mostly based on a classification key, which undertakes the choice of production process, the breakdown into job operations, and, on the basis of this, the creation of the work schedules. They are, however, in their use of traditional data processing solution algorithms, strictly limited by the sheer volume of possible alternatives. Here, an expert system might be able to solve the problem by limiting the number of possible solutions using heuristic methods. It duplicates the procedural approach of a human expert who, on the basis of his experience and specialist knowledge, can choose between a multiplicity of possible actions and alternative solutions which, in his opinion, have the greatest chance of success. Only if this is unsuccessful another strategy is adopted, another approach chosen.

The essential advantages of the use of such expert systems are:
- a heuristic approach to the choice of solutions, i.e. not all possible solutions are explored, only those with high probability of success (certainty factors);
- the problem area to be represented need not be fully described;
- they are capable of learning;
- they can explain the measures adopted and steps taken;
- they can requisition missing information needed for the problem solution;
- they can keep separate factual information, rules and problem-solving components.

An expert system (see Fig. C.VII.04) consists of 5 basic components (see *Harmon, King, Expertensysteme in der Praxis 1988; Scheer, Steinmann, Einführung in ES 1988; Puppe, Expertensysteme 1986; Schnupp, Leibrandt, Expertensysteme 1986*):

- the knowledge base (consisting of knowledge of facts and rules),
- a problem-solving or inferential component,
- an explanatory component,
- an interactive component,
- a knowledge-acquisition component.

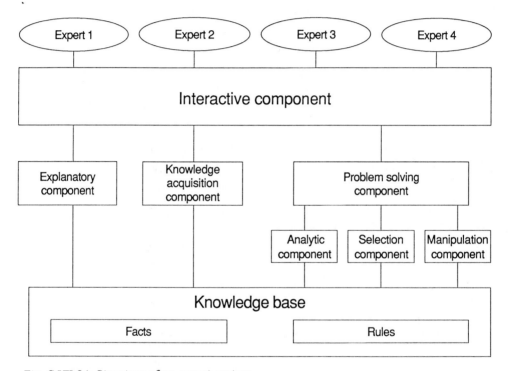

Fig. C.VII.04: Structure of an expert system

The individual components will be discussed using the expert system GUMMEX from the Battelle Institute as an example (see *Trum, Automatische Generierung von Arbeitsplänen 1986, pp. 69-73*). In GUMMEX work schedules for the production of elastomer products (rubber-elastic membranes for pneumatic use) are produced (see Fig. C.VII.05). Here emphasis is placed on the construction of the knowledge base, and structuring tasks are left to the general inferential mechanism (problem-solving component). The geometry information is obtained directly from the CAD system and prepared in a PROLOG file. The GUMMEX system currently contains about 400 rules and is available in a prototype system. It was written in PROLOG on a VAX-11/780.

146

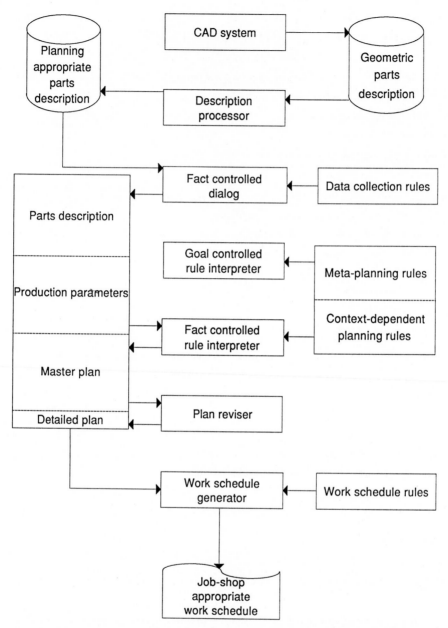

Fig. C.VII.05: Structure and integration concept of the expert system GUMMEX
from: *Trum, Automatische Generierung von Arbeitsplänen 1986*

The knowledge base (knowledge of rules and facts) represents the expert knowledge of the work schedule creator. Factual knowledge in this case consists of information about specific materials or technical parameters. Examples of rules might be (see *Trum, op. cit.*):

- Rule for determining the clamping order:
  - -- IF: (1)   the production part has 2 surfaces AND
    - (2)   both surfaces require precision processing AND
    - (3)   the tolerance between the two surfaces is very small
  - -- THEN: attempt to carry out the precision processing of both surfaces in the same clamping.

- Rule for choosing tools:
  - -- IF: (1)   the production part has a slot AND
    - (2)   the slot must be rough machined AND
    - (3)   the diameter D of the slot is greater than 50mm
  - -- THEN: (1)   the type of tool for the rough machining of the slot is a milling cutter AND
    - (2)   the diameter of the tool is less than D.

The problem-solving component searches for rules stored in the knowledge base with a high certainty factor (high probability of success), applies them and thus generates a solution space. The explanatory component can elucidate the chosen procedure and request missing information via the interactive component, which handles the communication with the user. These "new" rules or facts, not already stored in the system, are structured, characteristics and relationships are determined, and the "new knowledge" is entered into the knowledge component. For many tools a database link exists to store large quantities of data. The knowledge component is restricted to the current application area, whereas other components are applicable over a larger area. An expert system with an empty knowledge base is referred to as a shell. Shells created for medicinal purposes can, for example, also be used in technical units for diagnostic purposes. The fundamentally different procedural approach in the implementation of expert systems also requires different programming languages. Here, functional languages (LISP), relational languages (PROLOG) and object-oriented languages (SMALLTALK) may be used. Shells have now become available for all computer types in diverse price categories and with correspondingly diverse performance profiles (see *Rolle, ES für PC 1988; Rome, Uthmann, Diederich, KI-Workstation 1988*).

If attention is concentrated on order handling, the use of expert systems within CIM generates, on the basis of previously mentioned criteria, the following list of areas (see *Scheer, Steinmann, WBS in der PPS I 1989; Scheer, Steinmann WBS in der PPS II 1989; Steinmann, ES in der PPS unter CIM-Aspekten 1987*). Several expert systems which have already been developed in these areas are also listed (see *Mertens, Allgeyer, Däs, Betriebliche Expertensysteme 1986; Böhm, Konfiguration komplexer Produkte 1986;*

Krallmann, Expertensysteme für CIM 1986; Steinacker, Expertensysteme als Bindeglied zwischen CAD und CAM 1985; Krallmann, Anwendungen in CIM 1986):

- CIM handler - in sub-tasks for data and information management in complex systems,
- sales support for customer contact (XSEL),
- design support,
  - -- searching for similar parts and assemblies by accessing the PPC master files,
  - -- configuration of bills of materials,
  - -- through knowledge of design systems (e.g. taking new production and assembly procedures into account),
  - -- production and assembly suitable product design,
  - -- cost information at all stages of the design process,
- intelligent CAD systems,
  - -- design stage cost estimation,
  - -- product simulation,
  - -- configuration of end products from assembly variants according to customer specifications,
  - -- simulation of the planned production process (production techniques),
  - -- similarity planning,
- demand recognition,
- capacity planning, reconciliation of the planning levels,
  - -- master planning,
  - -- timetable (capacity planning and load levelling),
  - -- order release,
  - -- detailed scheduling/job shop control,
  - -- coordination of centralized and decentralized PPC systems,
  - -- evaluation of data from operational data collection,
- support for planning functions, PARERX, UMDEX, and DISPEX (see *Mertens, Borkowski, Geis, Expertensystem-Anwendungen 1988*), Panter (see *Steinmann, Konzeption zur Integration von WBS in PPS 1989*),
- capacity planning (simulation),
  - -- approaches to simulating the planned production process,
- order entry into production, short-term reservation planning (PEPS - Prototype Expert Priority Scheduler, or ISIS - Decision support system for job shop scheduling),
- process planning in production centers, flexible production systems, assembly islands (APLEX - Battelle Institute, Frankfurt, for work scheduling in the mechanical manufacturing area in processing centers),

- support for the creation of situation-dependent network plans (PROPEX- see *König, Hennicke, PROPEX 1987*),
- purchasing support,
  -- choice of suppliers (Purchasing Expert System) (see *Krallmann, EES 1986*),
- automatic work schedule creation by CAD (PROPLAN, GUMMEX, APLEX),
- generation of test plans and quality prescriptions by CAD,
- automatic NC program generation by CAD (Voest Alpine AG, Linz, among others, have a system in the implementation phase),
- error diagnosis in production stations (see *Schliep, ES zur Fehlerdiagnose an fahrerlosen Transportsystemen 1988*),
- robotics, linked with sensory and graphic processing,
- machine loading and process planning,
- production control, control systems for assessing the production process on the basis of code numbers (KOSYF),
- machine monitoring of CNC machines,
- error diagnosis for the servicing of production stations,
- material planning,
- material flow control and availability (ISA - DEC),
- diagnosis of PPC systems, similar to Siux (Siemens UDS-Expert),
  -- analysis and checking of stocks,
  -- analysis of throughput times,
- planning of the production sequence - plant layout
- maintenance, repair and servicing of the production facilities (FAULTFINDER, REPPLAN - Nixdorf), error diagnosis and repair instructions,
- servicing and repair of end products (DEX.C3 - an expert system for diagnosis of error behaviour in the automatic machinery C 3 at Ford).

In the computer industry expert systems are used for configuring plant and operating systems (Siconflex: used to configure the operating system of the Siemens SICOMP process computer from customer demands).

Digital Equipment Corporation (DEC) uses the following (see *Scown, Artificial Intelligence 1985*):
- XCON for system configuration,
- XSEL and XSITE for choosing system components according to customer requirements,
- ISA (Intelligent Scheduling Assistant) for determining production and delivery deadlines,

- IMACS (Intelligent Manufacturing Control System), XTEST (an intelligent system for finished product testing) and ILOG (Intelligent Logistic) for logical control of parts assembly.

As a result of the increasing use of sensors in production machines for recognition of production parts and checking of manufacturing tolerances, and scanners linked to automated conveyance and storage systems for object recognition, all the necessary data for automatic control and monitoring of the entire production area using an expert system are already available in computerized form. Since a system incorporating the total production area with all its elements, rules and interdependences would at present be too complex, it is necessary to start implementing expert systems in sub-areas, for example, the monitoring of production machines.

Even in the monitoring of production machines there is a relatively large event space due to the multiplicity of possible states and interdependences, such as tool locking, material shortage, changes in the good/bad parts count, and tolerance changes. The application area here, however, is already well-defined and capable of being structured such that factual and rule knowledge can be successively determined. With long term specification and storage of production tolerances, input of parameter combinations for identifying disturbances, filing of causes and necessary follow-up measures, all the decision prerequisites can be established and taken over as the knowledge base of the relevant expert system. Here, diagnostic expert systems are already in use and have proved their worth. Often quick decisions are essential in order to avoid damage to machines or products. On the other hand all unnecessary machine downtime is costly.

This aspect, too, was a motivating factor in introducing such systems. These aspects are of great significance, particularly in the case of production centers and islands . As a first step in this direction a control system was created on the basis of a production process model using KOSYF (see *Wiendahl, Lüssenhop, Basis eines Expertensystems 1986*). The next step is the establishment of an already pre-structured knowledge base. A large part of the factual knowledge can be taken directly from the existing database and then prepared. Even the basic structure of the rules is already established. Problems arise in coordinating expert systems with realtime requirements. In addition to the technical interfaces this requires logical levels defining how the system should react to incoming signals. For example, some signals require an immediate equipment shutdown, some signals require immediate checking of further system information, some signals require immediate notification of the operating personnel, etc. At present, the reconciliation of the logical and physical interfaces is still causing problems. In addition, there is the question of the competence of the expert system. Who carries the responsibility for defective decisions of the system, which in this context can have serious consequences?

After an initial period of excessive expectations from expert systems a phase of disenchantment follows and then realistic assessment of the application areas and possibilities. Already it is clear that the philosophy of expert systems can achieve enormous advantages in certain application areas. Successive extensions to existing systems and the growth of experience in using such systems will bring about a rapid increase in their complexity and efficiency. Expert systems will gain increasing importance in the CIM functional areas mentioned above.

2. Expert System for Design Stage Cost Estimation

(*Dipl.-Inform. Martina Bock, Dipl.-Inform. Richard Bock, Institut für Wirtschaftsinformatik (IWi), University of Saarbrücken*)

2.1 Introduction

Design stage cost estimation can be broken down into the sub-tasks "preparation of cost information" and "cost calculation". In the preparation of cost information context data can be made available in the form of cost values or as relative costs and knowledge about the kind of effect of factors relevant to costs. In the cost calculation context many diverse procedures are well known (see *Scheer, Konstruktionsbegleitende Kalkulation 1985*).

These can be distinguished on the basis of the fundamental methodology, the data requirements, the accuracy of the results, and their suitability for various design problems (see *Ahlers, Gröner, Mattheis, Konstruktionsbegleitende Kalkulation im Rahmen des CAD 1986*). However, these systems are often tailored to a small problem area, such that they are only applicable to special products and operations. A generally applicable, "intelligent" system for design stage cost estimation needs to fulfil the following requirements:

-   Preparation and application of specialized and general knowledge from design and cost estimation:

    The design knowledge serves to pass on the knowledge of an experienced designer. Cost estimation knowledge provides the designer with data, methods and procedures from cost accounting. The essential aims are the determination of cost relevant factors and the choice of a suitable accounting procedure for the case in question. In addition, the knowledge should be applied to provide or generate facts about a product which allow accurate accounting procedures to be used as early as possible.

-   Applicability throughout the entire design process:

    Throughout the various design phases data of varying accuracy need to be accessed. Data which are not specified need to be derived.

-   Provision of active assistance:

    The aim is to interpret the results of the accounting procedures used and to ensure situation-specific cost optimization.

-   Flexibility of application:

    The system should be easily adaptable to product-specific and organizational situations.

A prerequisite for the development of a system satisfying the requirements described is the use of methods and procedures familiar from the artificial intelligence area, and especially the development of knowledge based systems.

2.2 Structure and Mode of Operation of Expert Systems

The task of the expert system is to provide the designer with cost information during the first three phases of the design process. This assumes a sub-division of the design process in accordance with VDI-Guideline 2222 into planning, conception, outline and detailing phases. The system must have knowledge of the data available in each phase, be able to create links with the accounting procedures based on this information, and to derive information which is not available. On the basis of these tasks a design phase-oriented mode of operation and processing strategy of the expert system results.

2.2.1 Design Stage Cost Estimation in the Planning Phase

Before cost data can be established in the planning phase of the design process, it is necessary to locate a similar product group on the basis of the requirements list for the new product and the requirements (physical characteristics) of products that have already been designed and produced. This similarity search presumes that a classification of products according to physical characteristics has been undertaken and that products with the same basic bill of materials structure have been combined into product groups. In this context the basic bill of materials structure contains only those parts of a bill of materials which reflect the fundamental structure or assembly of a product without consideration of optional parts. For example, according to this definition, an optional second wing mirror on a car would not belong in the basic bill of materials.

All those design alternatives which have already been generated in the firm can be represented in a hierarchy (see Fig. C.VII.06) differentiated according to the dominant product and cost determining factors and special operational conditions. This hierarchy, together with the requirements list for a new product, provides the basis for the similarity comparison.

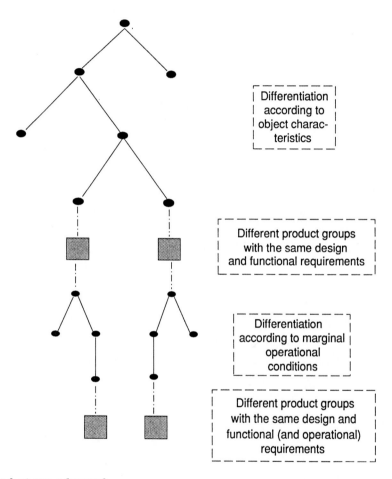

Fig. C.VII.06: Product group hierarchy

Within the similarity comparison special design rules serve to recognize and take into account design dependences between sets of physical characteristics.

An example of a design rule might be:

IF: a special surface treatment is desired,
THEN: the material used must be suitable for this treatment.

After a successful similarity comparison, which for the first phase implies that the product to be newly designed can be regarded as a variant of an already designed product, the choice and application of flat-rate cost estimation procedures can be undertaken (see *Scheer, Konstruktionsbegleitende Kalkulation 1985*). If several procedures can be applied, they are prioritised on the basis of which can provide the most accurate cost data. A hierarchy of cost estimation procedures is used to determine which cost estimation procedures could be applied. The specification of the estimation formulae, which often requires extensive statistical evaluations, is not supported by the expert system. Since the use of flat-rate cost estimation procedures does not in general lead to very exact cost data, a further step attempts to generate the precise geometric dimensions of the product either automatically or in interaction with the designer, such that more precise cost estimation procedures can then be applied. The basis for automatic generation are the basic bill of materials for the most similar product group and some generation rules.

An example of a generation rule might be:

IF: the required loading capacity has the value x,

THEN: the strength of the surface, dependent on its length and width, must have the
         value y.

Successful generation of the geometric dimensions then allows the application of cost estimation procedures based on geometry data. The cost of variants which only differ from the basic bill of materials in that parts are added to the product which have no effect on the basic structure of the product (e.g. the second wing mirror on a car) must then also be established and added to the costs determined by the cost estimation procedure. In addition to providing a similar basic bill of materials for a product to be newly designed, a successful similarity comparison also automatically provides a similar work schedule (the work schedule of the most similar product). By modifying this work schedule in accordance with the geometric dimensions generated for a product the exact production costs of the product can be calculated. The work schedule can be modified either using heuristic rules or in interaction with the expert.

The processing steps described thus far make it possible, simply on the basis of a requirements list for a product, to obtain in the first design phase those data which are normally only established in the subsequent design phases, and to use them to calculate the cost of the product. A prerequisite for expert system assistance of this form in the first design phase is that the product can be represented as a variant of an already designed product.

The data obtained also provide a suitable basis for cost optimization. With the help of a material classification system, in which raw materials are classified according to technical, physical characteristics, the costs of using alternative raw materials can be calculated. The optimization of production costs implies either finding a lower cost variant work schedule or replacing a production process with a cheaper one. To do this production processes must be classified in a similar way to raw materials. The entire optimization process is conducted automatically.

2.2.2 Design Stage Cost Estimation in Phase 2 of the Design Process

If the similarity comparison in the first phase of the design process was unsuccessful (i.e. there are no variants), an attempt is made in the second phase to assign the functions of the product to solution principles and assemblies. The basis for this is a classification of all the functions already existing in the firm and their assignment to assemblies and individual parts. If a similarity search is successful here an adjustment design is available. Once the geometric dimensions of the identified assemblies and individual parts have been determined, either using rules or interactively with the designer, an initial estimation can be made for the product in the form of function costs or costs of individual assemblies or parts. The subsequent processing steps, which are represented in Fig. C.VII.07, are executed in a similar fashion to those in phase 1.

156

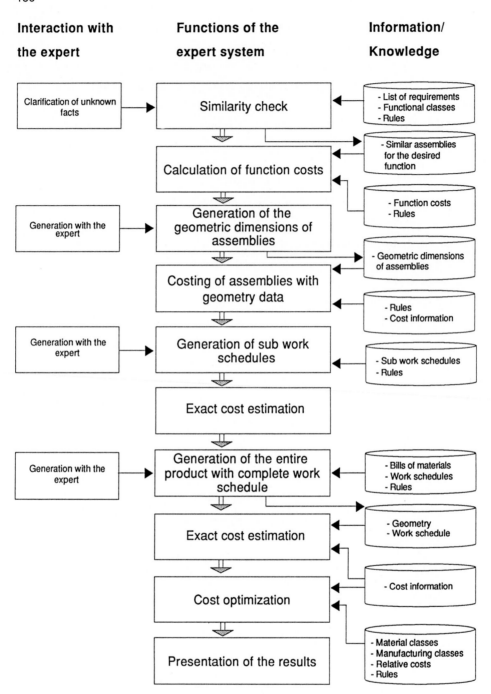

Fig. C.VII.07: Design stage cost estimation in phase 2

157

2.2.3 Design Stage Cost Estimation in Phase 3 of the Design Process

If satisfactory cost predictions are not possible in either phase 1 or phase 2, then the product to be designed is a genuine new design. Before costs can be determined in this case, the designer must determine the geometric dimensions of the product. Thereafter, determination of costs can be made with the help of cost estimation procedures based on geometry data. Basically, for new design it is the case that recourse to existing data is scarcely possible. Precise determination of costs, without being able to access already existing work schedules, for instance, therefore requires a great deal of very general and product-specific knowledge from the cost estimation, design, work scheduling and production areas. Consequently, it is only possible in very limited cases.

2.3 System Architecture

The system architecture outlined in Fig. C.VII.08 results from the requirements and tasks of the expert system that have been described.

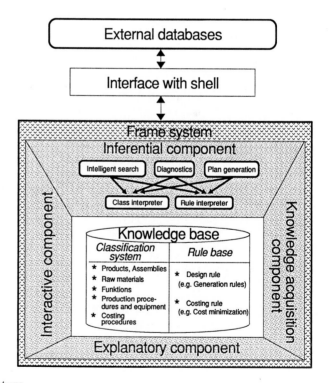

Fig. C.VII.08: System architecture

The various classifications of products, raw materials, cost estimation procedures, functions and production processes are stored in object-oriented form together with the various types of rules in the system's knowledge base. The general structures and data are pre-determined, whereas the product and organization-specific data and rules are generated interactively with the expert in the course of the knowledge acquisition or are taken from the organizational database. A program link translates the data from the organizational database into the object-oriented form. In addition, it evaluates queries to the organizational database and the cost information system relating to the run-time of the expert system. The design phase-oriented processing of the various steps is controlled by the inferential component, whereby necessary interactions with the expert during the system run time are coordinated by an interactive component. A link between the expert system and a CAD system is envisaged to present results and access geometry data.

VIII. Computer-Technical Model

The computer-technical model comprises the following areas:
- hardware,
- network structure,
- system software,
- database systems,
- application tools (CASE, data dictionary).

For CIM, establishing the hardware architecture is of particular significance, since this involves the linking of diverse systems as well as the support for the functional architecture derived above. This point will, therefore, be discussed in its own right. The other issues will be handled alongside the fundamental possibilities for system integration.

a. Hardware Architecture

On the right side of Fig. C.VIII.01 a detailed functional architecture for an industrial firm is represented, similar to that of Fig. C.V.01.

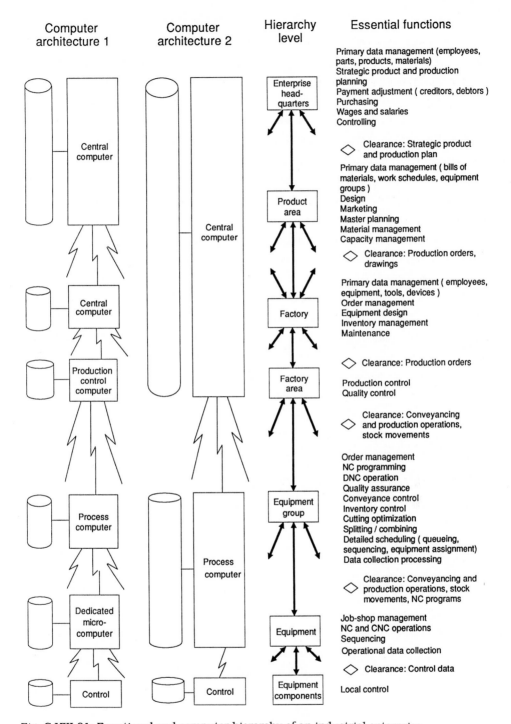

Fig. C.VIII.01: Functional and computer hierarchy of an industrial enterprise

This functional assignment is the basis for the hardware architecture to be created. Two possibilities are represented in the figure. This will make it clear once again that the number of hierarchy levels needed for definition of functions is not necessarily the same as the number of computer levels.

In the two level model of the example, functions from the company headquarters to the factory area level are executed on a general purpose computer. The reason for this might be that all four levels are physically located in the same place and the customer order management data needs to be very up-to-date.

For the more technically-oriented areas within the factory area a second computer level is depicted. This computer level also directly serves the lower level computer controlled production processes.

Because of its involvement with technical systems the second computer level is treated as a process computer. Hardware and operating systems make it possible to use various peripheral devices as well as particularly suitable operating systems (e.g. with realtime capabilities). The de-coupled computer support also ensures the availability of data processing support independent of the higher level general purpose computer. By holding appropriate order buffers it is even possible for the night shift to obtain data processing support when the host computer is not available.

The architecture on the left of Fig. C.VIII.01 provides a greater spread of computer levels. Here, a general purpose computer is also installed at the factory level so that it is easier to create links with the product group which deals with customer order management.

To carry out the functions of the factory area this level is intended as the production control computer. Below this level process computers are installed to handle the more technical functions.

The formation of several levels is based on criteria such as the assurance of high availability, required response time characteristics, or reduced complexity as a result of stratification of tasks.

In choosing concrete hardware systems for the computer architecture that has been established it is also necessary to take account of integration requirements. Basically, at the moment it is still not possible to link up whatever (heterogeneous) hardware systems such that they can communicate with each other.

For this reason the large hardware manufacturers have developed within their product lines their own self-referenced integration models. For instance, the IBM system SAA (System Application Architecture) offers a model which ensures that the applications software developed is capable of running on diverse types of computer. If the computer hierarchy is implemented using SAA components (such as database systems, network concepts, programming languages, Case Tool, etc.) a high level of integration can be ensured. A similar model is provided by Digital Equipment via the operating system VMS, the database system RDB and the network concept DECNET which are available on all its products.

It should be noted, however, that adopting such a manufacturer's model limits the choice of products. As a result situations can arise where essential requirements, resulting from the use of applications software or special dedicated hardware systems, cannot be supported.

Consequently, instead of manufacturer-specific models the "open standards" concept can be adopted. This strategy provides greater independence from manufacturers. Typical standards here are UNIX as operating system, X-windows and MOTIF as user interface, SQL as database interface, the programming language C and MAP as network concept. Leading manufacturers have got together to form organizations to ensure uniform definitions of these products (e.g. OSF (Open System Foundation)). Of course, the manufacturers do not all offer an identical product, but rather derivatives of the standard. Nevertheless, it remains the case that this development has brought about greater coherence, and integration can be achieved more easily even between derivatives than is possible between manufacturers' own developments.

In implementing a hardware concept the enterprise cannot generally make a decision in complete freedom. Rather, existing structures have to be taken into account. Since computer support in the production area is still relatively meagre, however, this presents a good opportunity to create a flexible basis for the future by adopting a forward-looking computer strategy. It is, therefore, a sensible strategy to continue to use the currently existing hardware at the host level, but to begin to adopt a new computer strategy at the factory area and equipment group level in which, for example, open standards are rigorously implemented.

Further hardware aspects will be considered in the following section.

b. Integration Instruments

In order to achieve the data and applications integration between the CIM sub-components required to support closed process chains, not only hardware integration, but also the linking of operating systems, database systems and application interactions is necessary.

However, many hardware producers and software houses have only specialized in the left fork of Fig. A.01, that is, in the more commercially-relevant information processing. On the other hand producers of CAD and CAM systems have little experience with more business-oriented data processing. This has given rise to systems which are exclusively oriented to the isolated functions. Since firms in the user community also often display a separation of business and technical departments, recognized interdependence between the systems tends rather to be under-emphasized, and not to be a criterion for choosing between systems. Since the technical areas still possess considerable freedom in investment decisions, CAD systems are typically chosen simply on the basis of their functional qualities. Here again, the suppliers are encouraged in their policy of specialization.

This specialization has led to a situation in which the user employs a multiplicity of diverse systems to support his process chains (see Fig. C.VIII.02).

Process chain	Software
Product concept	CADAM and simulation software
Configuration	Flight behaviour and sizing
Aerodynamics and	Software for solving difference
aeroplastics	and differential equations
Strake design	Geometric processors, such as G3D
Preliminary design	CADAM
Stability testing	Finite elements, COSA-DEMEL
Design	CADAM, COPICS
NC programming	CADAM extensions APT and APL
Production organization	COPICS
Production	DNC (direct numeric control)
	Production data collection

Fig. C.VIII.02: Process chain
from: *Sock, Nagel, CAD/CAM-Integration 1986*

As a result of the recognition that more growth is to be expected in the technical information system area then in the commercial area, suppliers of commercial systems are beginning to develop an interest in technical data processing applications. Conversely, since the rise of CIM the more technically-oriented producers have had to offer commercial systems. The manufacturers, however, cannot build up competence in the other area from scratch and develop systems which can be integrated with the software available up to now to form a CIM system.

It is therefore understandable that hardware and software manufacturers only gradually succeed in achieving integration from the starting point of their isolated systems. Fig.

C.VIII.03 represents the different levels of computer-technical integration. The two CIM components CAD/CAM and PPC are presented as an example of two systems to be linked. Other components, such as CAQ and process planning from CAP, or the link

1. Level: Organizational integration of independent data
   processing systems

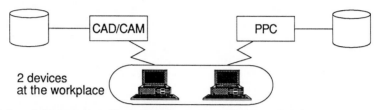

2. Level: Integration of independent systems by way of tools
   (PCs, Query, Networks)

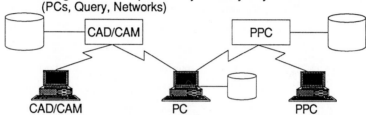

3. Level: Data transfer between systems

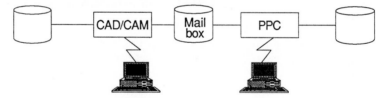

4. Level: Systems sharing common database

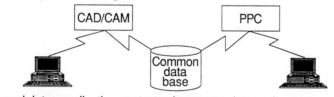

5. Level: Inter-application contacts via program integration

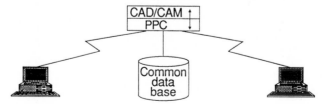

Fig. C.VIII.03: Integration levels and possibilities

between material management and a dedicated production control system within PPC, could equally well have been chosen.

The first three integration levels represent concepts that are at present widely available, whereas the fourth level is only available in special cases, and the fifth level is not yet on offer, although clear efforts towards its realization are discernible on the part of software and hardware manufacturers.

1. Organizational Solution

The first level merely represents an organizational link between computer-technically independent CAD/CAM and PPC systems. This means that personnel in material management, process planning or design are equipped with two displays, with whose help the relevant systems can be accessed. Information functions can be carried out, but data cannot be automatically transferred from one system to the other; this requires manual transfer. As a result data consistency between the various databases can in no way be supported by computer-technical means. This level, therefore, can only be seen as a makeshift integration solution.

This integration level is independent of whether it is effected via a purely organizational combining of separate systems or whether the separate systems are accessed from a terminal (terminal emulation). In both cases, however, there is no data flow between the systems.

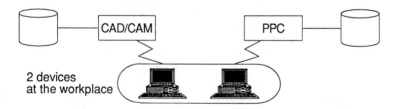

Fig. C.VIII.03,a: Level 1: Organizational integration of computer-technically independent systems

2. Use of Tools

In the second level framework the basic CAD/CAM and PPC systems are still operated independently, but are linked using data processing tools, which thereby represent integration in the "third dimension". This integration has the advantage that the basic

CAD/CAM and PPC systems remain unchanged. Although exploitation beyond the scope of either system is now possible, the disadvantages of inadequate support of data integrity remain. Along with the use of microcomputers (sometimes on the integration), database system query languages and LANs can be implemented to handle this task. Given their obvious importance for CIM implementation these will be considered further.

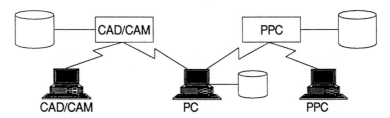

Fig. C.VIII.03,b: Level 2: Integration of independent systems using tools (PC, query, networks)

2.1 Microcomputers/Workstations

Since microcomputers (personal computers, workstations) have from the start been conceived as open systems, they can be linked up to a variety of hardware systems, including independent hardware systems for planning and technical processes. In the meantime almost all manufacturers have oriented themselves to the so-called "open" and/or "industrial standards". Around these standards a great number of producers of highly differentiated hardware extensions, communication facilities and peripheral devices has grown up. Hence at this hardware level there numerous solutions to practically every linkage problem between computers of the most diverse types exist. By simultaneous access to both systems and the use of high performance query languages, file transfers from the CAD/CAM or PPC system into the PC database can be carried out following a selection procedure. The assembled data are then available on the PC for integrated evaluation, using comfortable evaluation systems (spreadsheet software).

The introduction of more powerful PC operating systems such as UNIX or MS-DOS 5.0 largely eliminates the restrictions of the strictly limited address space available in PC applications up to now (see *op. cit., MS-DOS 5.0 1986*). 32-bit architectures are being supported, and addressable central core extended to 16 MB. For the integration tools mentioned in the context of a CIM concept using PCs, it is precisely this aspect that is of great significance. Window techniques, networking possibilities and the availability of modern database systems are of primary importance in the use of PCs as CIM integration tools. **Window techniques** allow the user to access several applications simultaneously on the screen, whereby each application is assigned a "window", i.e. a particular section of the screen (see Fig. C.VIII.04). These windows can be overlapped and changed in size. The changeover from one window to another enables the user to coordinate the **simultaneous utilization of several functions** at the workplace. This can be effected without the time-wasting process needed up to now of logging out of one system and logging into the other. What is more, information from several applications is simultaneously available on the screen and this facilitates efficient, inter-functional work methods. A work scheduler could in this way have simultaneous access to drawings, bills of materials and equipment cost data on the screen.

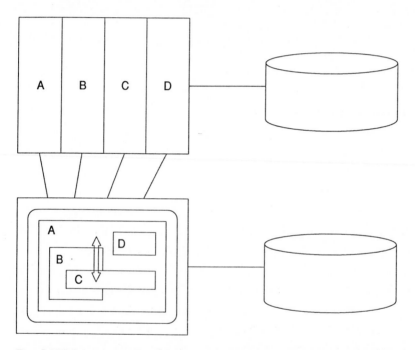

Fig. C.VIII.04: Integration of different applications using window techniques

The **linkage of different application systems** is possible using window techniques, where the user initiates a data transfer from one application window to another. Messages or status information from current adjustments can thus be taken account of immediately in the applications of other functional areas of the organization.

Along with the use of PCs as terminal emulators incorporating window techniques, the network capabilities of modern PC operating systems allow the possibility of effecting inter-functional message processing in the context of CIM chains using PC mailing systems. These procedures allow the PC to take over the normal terminal functions of the employee's workplace. The window technique encourages a functionally integrated approach and, finally, PC networking provides a cost effective inter-functional communication chain.

As well as allowing parallel access to several functions via the window technique, the PC can also be usefully integrated with existing (mainframe) applications as a query or evaluation instrument. PC database systems are often every bit as powerful in their functional scope as large computer systems, and even their data format is often compatible with large systems (ORACLE) (see *Scheer, Aufgabenverteilung Mikro-Mainframe 1985, p. 5*). PCs can, therefore, contribute to increased availability of computer capacity, in that they can perform analysis and evaluation functions by accessing extracts from the host data records. Such applications might be simulation functions in the master planning context, or detailed scheduling or sequencing in the production control context. With reference to the production island concept, the speed of access can be increased by providing each PC locally with a specific subset of the central PPC data; amendments to the original data are then updated by the host machine (see *Kernler, Einsatzspektrum des PCs für PPS-Aufgaben 1985*).

2.2 Database Query

Along with the information and evaluation functions provided in application programs, the use of new software technology makes easily-learned, user-friendly query languages available to the user. The provision of such tools is a significant indication of the quality of a database system; here relational database systems stand out, particularly for their user comfort.

Database query languages are offered by many database manufacturers. Examples are QueryDL/1 for DL/1 (IBM), SQL for DB2 (IBM), SQL for ORACLE (Oracle), NATURAL for ADABAS (Software AG) or SESAM-DRIVE for SESAM (Siemens).

With the help of a basic structure using only three key words, SQL allows quite complex questions to be framed:

SELECT	Fields (columns) to be displayed,
FROM	Relations (files) containing the fields,
WHERE	Conditions, giving the selection criteria.

Query - also called Very High Level Language (VHL) - has proved to be an excellent integration tool for the implementation of CIM, because it allows inter-system applications and increased flexibility.

If the various operational sub-systems provide a unified user interface in query form, the user can initiate inter-system queries, e.g.

- job-specific progress control of a customer order (PPC and CAM),
- on-line credit checking with up-to-date financial and order data (PPC and the financial systems).

By using query the user can independently access information necessary for his area. Hence his applications are independent of the data processing department, which normally provides the user with information and evaluations in the form of prescribed programs using menu techniques. In view of the overburdened data processing departments and the consequent build-up of applications for new functions, the user can obtain his particular information more quickly and flexibly.

Because of their special requirements, geometry and drawing data from the CAD area tend to be less suited to storage in relational databases. In order to be able to store all operational data in a unified database, therefore, CAD manufacturers are developing new approaches, in which geometry and drawing data are removed from the CAD database and translated into a neutral format, provided with descriptive parameters and placed in a relational database. With the help of a query language the user can now access the descriptive parameters of the geometry and drawing data within an administrative application.

2.3 Networking of Computer Systems

If computers are linked together such that data can be exchanged between them this is referred to as a computer network. Where the physical distances are larger the public network services of the German Bundespost (HfD, DATEX-L, DATEX-P) are used.

Computers are linked together (see *Schnupp, Rechnernetze 1982, p. 17*) in order to

- even out capacity peaks (capacity network),
- make special hardware available in several places without it needing to be physically located there (equipment network),
- make data stored in diverse locations available to other locations (data network),

- coordinate several sub-tasks assigned to individual computers into a single overall task (intelligence network),
- provide postal-type services for communication between physically distant human users (Electronic Mail, Electronic Conferencing) via computer systems (communication network).

Of these network forms, data networks, intelligence networks (often linked with equipment networks) and communication networks are of the greatest economic significance, although in specific applications several network forms are generally implemented together.

2.3.1 Types of Network

Technical network implementations can be distinguished according to their
- topology,
- transfer medium and
- transfer protocol (network access rules).

As regards the **topology** of computer network systems four possibilities can be distinguished (see Fig. C.VIII.05).

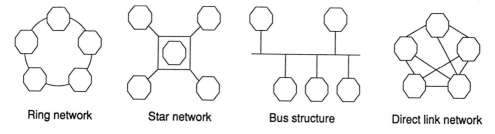

| Ring network | Star network | Bus structure | Direct link network |

Fig. C.VIII.05: Computer network topologies

In a ring network all nodes have equal status, i.e. the network can be controlled by any node. Each station is "active", since a message can be checked by any station on its route, irrespective of whether the message is being received there or merely passed on. If one station is down the entire network is interrupted. On the other hand, the network is easily extendible and leads to low costs for cabling and the required protocols. It is therefore also used for PC networks.

An unambiguous hierarchy is characteristic of star networks, whereby network control is effected by the central node. This topology is typical of terminal networks in which many terminals are linked to a host computer. If one terminal is down this has no effect on the other stations, but if the host computer is down the entire network fails. Disadvantages of this form are the limitations to extendibility by the performance capabilities of the host and the high cabling costs.

Large quantities of data can be transported on a bus cable. The stations, however, have no control functions, they are "passive". As a result, if one station is down this has no effect on the others. As a result of this insusceptibility to breakdown and the ease of extendability for relatively low cabling costs the bus topology has become widespread. In particular, it has become predominant for local area networks (LAN).

LANs are networks with a very high throughput rate, to which the most diverse hardware systems, from diverse producers, with diverse interfaces such as terminals, printers, microprocessors, plotters and also mainframe computers can be connected. They are restricted to some limited physical space, e.g. a building. They have the advantage that, for networks of this type the monopoly restrictions of the Bundespost for network and transfer structures do not need to be adhered to.

LANs are of particular importance in connection with office automation, since within spatially limited organizations text processing, data processing, graphical output and word processing can be linked with each other.

In a network with direct links there may be a control center, but the network links need not necessarily run via this center.

The reason for installing a network with direct links, and hence with redundant data paths, is greater protection against breakdown. A star network, for instance, has only one path between two points, a ring network has two, a network with direct links has several paths, so that if one route is down there is a variety of alternatives.

As regards the **transfer media** a distinction is drawn in networks between broadband and baseband cables. A broadband cable has a high throughput capacity. The frequency range can be subdivided into various areas (channels), each of which handle separate links. Broadband cable is suitable for high transfer speeds (48,000 baud; baud = bit per sec.) and is implemented with glass fibres in new technology. Since video transfers are also possible, they are also discussed in the office application context.

Baseband cable is currently used principally for telephone links. It has a transfer rate of 4,800 or 9,600 baud and is therefore suitable for moderate data transfer requirements.

For network accessing (especially for LANs) two **transfer protocols**, CSMA/CD (Carrier Sense Multiple Access with Collision Detection) and token passing, are of particular significance.

In CSMA/CD a station "listens" into the network to see if it is free for a transfer, i.e. whether any other message is currently being handled. If the network is free the station then transmits. This can give rise to "collisions" if another station transmits at the same moment. In this case a random algorithm determines the network access sequence. This accessing principle was implemented in the ETHERNET network system developed by XEROX, and has also been adopted by other suppliers such as DEC (DECNET).

In the "token passing" principle a bit pattern (token) wanders around the network. A station wishing to transmit waits for a free token and then attaches the message to be sent to it. The receiving station removes the message and the token is then free again. Since this principle implies that a message can only be sent when the network is free there is no danger of collision. The token passing procedure is offered by IBM, especially for PC networks in token ring form.

To facilitate the linking together of different stations, networks adopt protocol standards (see *Schäfer, Technische Grundlagen der lokalen Netze 1986*). If these are supported by the hardware systems of different manufacturers then even heterogeneous hardware units can communicate with each other.

The most commonly employed industrial standard is the TCP/IP family (Transmission Control Protocol/Internet Protocol) (see *Blum, Engelkamp, Porten, Kommunikationsnetze für CIM 1988*). TCP, which is built on IP, provides secure data transmission between two systems. IP takes over the transport of data packets from one sender via several networks to one recipient, whereby larger packets are broken down into smaller parts and routing decisions taken. For applications the File Transfer Protocol (FTP), which undertakes file transfer and program call up on remote machines, and the terminal protocol TELNET, which offers communication between terminals and interactive applications processes, may be employed, for example.

In the long term it is expected that internationally standardized protocols, such as the ISO/OSI reference model or the IEEE standards for example, will be of much greater importance (see *Haberstroh, Nölscher, Im Netz von CIM 1989*). In this context MAP (Manufacturing Automation Protocol), which has been developed for manufacturing applications, is receiving particular attention.

2.3.2 The Aims of MAP

The integration of material flow control using computer controlled storage and conveyance systems, automated manufacturing installations (NC, CNC, DNC, Robotics) and the regulating functions of a control center computer demands that the various computer controlled systems can communicate with each other. The present situation is characterized rather by incompatibilities. Each computing and control system possesses its own code for drawing representation, its own formats for records and files, its own protection mechanisms and its own definitions for the control of peripheral devices. Sometimes the systems are embedded in extensive operating system environments, which are also manufacturer-specific. Generally, when manufacturers offer their own network concept (e.g. IBM with SNA, Digital Equipment with DECNET, Siemens with SINET and TRANSDATA, Allen Bradley with DATA HIGHWAY or Gould with MOTBUS) only devices of the network manufacturer are useable (see *Suppan-Borowka, Anforderungen an MAP 1986*).

At the end of the 1970s General Motors (GM), as the largest enterprise in the world, began its own efforts to develop a standard for networking in the factory automation context. The "Manufacturing Automation Task Force" was founded by GM, which, between 1980 and 1983, designed the concept of an open communication system MAP (Manufacturing Automation Protocol). As a result of its market power, GM then succeeded in involving a range of important manufacturers of information technology products in the project. In the meantime the MAP concept has been defined to such a stage that, in addition to several test installations such as have been demonstrated at the Hannover Show or Autofact, it has been possible to develop concrete MAP products. Meanwhile, in Europe institutions such as the European Map Users Group (EMUG) have developed, which reveal the strength of interest of European manufacturers. Since MAP presents itself as an industry standard, all leading producers of information technology components for manufacturing (producers of controls and OEM partners who build controls into their production installations) are now coming to terms with this standard.

For the user the opportunity is developing to employ unified communication paths to join together a multiplicity of EDP systems, a situation which cannot be avoided, due to the complexity of the problems posed in manufacturing. This in itself represents an enormous economic advantage, since the cabling costs for non-standard transmission media, each of which requires its own links, not only involve an immense cost in terms of individual protocol matching, but also lead to chaos in the laying of cable, with overflowing cable ducts. The costs of cabling itself can constitute a considerable proportion of automation costs - estimates put this at 50% to 70% of the total cost.

An impression of the immensity of the problem at GM is given by the fact that around 40,000 intelligent EDP systems were in use, including 20,000 programmable controls and 3,000 industrial robots. Of these 40,000 systems only 15% could communicate with other systems (see *Neckermann, Das Netz von morgen 1985*).

In developing MAP, existing standards should be used as far as possible, and only when these provisions are lacking additional definitions should be introduced. The basis is, therefore, the ISO/OSI reference model for open systems (see Fig. C.VIII.06).

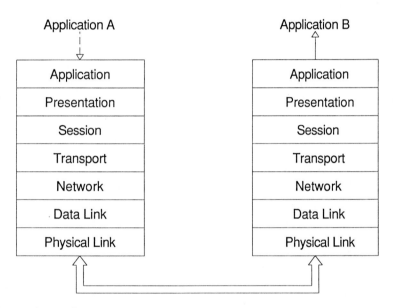

Fig. C.VIII.06: The ISO/OSI reference model

2.3.3 MAP Functions

In order that application A can communicate with application B the standard service calls up the uppermost layer (**layer 7**) of the communication system (see Fig. C.VIII.06). The data are then consecutively transferred to subordinate layers, and supplemented with additional control information, so that they can be distributed from the first layer via a physical transfer medium to other participants. From here the data travel from bottom to top through the layers, with each layer taking and interpreting the corresponding control data from the parallel layer (sender, recipient). Finally, those data that are relevant to the proposed application B are transferred.

Applications which can be included in this schema are basically all the organizational and technical applications, such as bookkeeping, cost accounting, production planning,

CAD, office communication, statistics, etc. However, at the first step MAP concentrates on the exchange of NC control programs. At this layer, significant decisions need to be made in the choice between several possible standardization options. Fig. C.VIII.07 shows the individual tasks, as described by the ISO reference model, set against the respective MAP specifications. The four lower layers are nowadays generally recognized and accepted, whereas the higher layers are still subject to discussion. These discussions have given rise to the development of different versions. Fig. C.VIII.07 shows, in addition to Version 2.0, the Version 3.0, which was first shown in Europe at SYSTEC 88. Version 3.0 is regarded as pathfinding, to the extent that all significant computer manufacturers have announced concrete MAP products.

ISO-Layer	Task	MAP-specification 2.0	MAP-specification 3.0
Layer 7 : Application	Service interface for users	MMFS/EIA 1393 A ISO-CASE-subset ISO-FTAM-subset MAP-messaging MAP-directory-service MAP-network-management	MMS/RS - 511 Filetransfer (FTAM) Network management (NM) Directory services Virtual terminal CASE
Layer 6 : Presentation	Conversion/amendment of formats, code, etc.	Null	ISO-presentation -service -protocol
Layer 5 : Session	Link synchronization and management	ISO-session-kernel	ISO-connection oriented session -service -protocol
Layer 4 : Transport	Reliable end-to-end links	ISO-transport-class 4	ISO-transport -service -protocol
Layer 3 : Network	Protocol coordination between different networks, routing	Currently null	ISO-INTERNET connectionless
Layer 2 : Data Link	Error discovery, transfer between topologically adjacent nodes, media access	IEEE 802.2 LLC Type 1 IEEE 802.4 tokenbus	LLC Typ 1 802.4 tokenbus
Layer 1 : Physical Link	Coding and bit-series transfer of packages	IEEE 802.4 tokenbus-broadband	Broadband, carrier band, fiber optic ISO-tokenbus 802

Fig. C.VIII.07: Communication in automated production

A fundamental distinction needs to be drawn between the three lower layers and the higher layers of the ISO model. In the three lower layers the physical transfer of data is conducted in response to messages, whereas the higher layers have a closer relationship with the applications. Given the multiplicity of possibilities standardization of applications requirements has proved exceedingly difficult. This is also indicated by the different versions of MAP. Decisions taken at layer 1 must limit the number of different access procedures and, related to this, different transfer media.

Comparing the ISO model with the industrial standard TCP/IP it becomes clear that TCP belongs to layer 4, IP to layer 3 and the protocols TELNET and FTP to the layers 5 - 7.

Fig. C.VIII.07 depicts only the basic recommendations; MAP also supports further possibilities, or at least does not exclude them. In spite of this the so-called token bus with a broadband medium in accordance with standard IEEE 802.4 can be regarded as MAP standard for layer 1. The broadband system has a transfer rate of 10 MBit/sec. It supports transfer over large distances, the parallel transfer of several information streams, and possesses an adequately large transfer capacity.

The special properties of MAP 3.0 as compared with the previous versions can be seen at layers 6 and 7. By implementing layer 6 in version 3.0 diverse formats can be adapted. At the heart of MAP, however, lies the definition of duties at **layer 7**, which is directly informed by the higher layer applications. Here it is a question of the definition of file transfers, network management, directory services, and of virtual terminals, which are a prerequisite for the link up of various devices. The necessary functions are laid down by ISO in specific rules such as MMS - the "Manufacturing Message Specification" (previously RS-511) or the "Abstract Syntax Notation Number One" (ASN.1) (see *Gora, MAP 1986, p. 40 ff.*). MMS is a standardized language for data exchange in the production environment. In order to control CNC and data collection systems, for example, functions such as the down-loading and up-loading of programs, queries as to device status, error messages and the call-up of program functions are made ready. MMS defines data, objects, and messages for carrying out specific operations in an abstract form so that, in

Function	Description
CYCLE START	Activation or termination of the current machine cycle
PART	Identification of individual work-pieces
AXIS OFFSET	Axis manipulation (e.g. for robot control)
UNITS	Definition, which units (inches/meters) are to be used
EXCHANGE	Manipulation of work-piece palettes
LIFT	Lifting of a specific instrument

Fig. C.VIII.08: MAP instructions
from: *Gora, MAP 1986, p. 41*

contrast to the earlier version MMFS (Manufacturing Message Format Standard), it is device-independent. Fig. C.VIII.08 gives an impression of the individually defined instructions for a numerically controlled production installation.

Economic efficiency aspects are linked with the development of MMS. For example, non-recurring, system-independent program creation generates cost savings; the use of standard software is also conceivable. Furthermore, applications can be allocated to the most suitable systems depending on the task and the load.

The basis for the definition of layer 7 is the ISO protocol CASE (Common Application Service Element). It is extended by a range of functions for the exchange of NC data. Here, the Application Layer Interface (ALI) of Version 3.0 offers particularly user-friendly access to the network services.

A general control system for the network is also defined in MAP. Constituents of this function are the monitoring and specification of the network configuration, as well as monitoring and the remedying of faults. Since faults can arise at any layer, layers 3 - 7 are covered by MAP within the management concept. For layers 1 and 2 suitable security procedures are already available in the definitions taken over from ISO.

Although the MAP architecture allows priority control using the token bus system, this is not implemented, so that realtime application is not yet available. For current applications in transferring NC control data, particularly for file transfers, this is not needed. Version 3.0 achieves performances in the 20 - 50 msec. range, so that this already comes very close to realtime operation.

2.3.4 Embedding of MAP in General Network Architectures

Although MAP represents a spectacular step in the standardization direction, at least in the case of manufacturing related LANs, it does not fulfil all the requirements of an open network architecture in an industrial firm. On the one hand, firms have already introduced sub-networks linking automated islands, which have to remain at least in the medium term, and therefore need to be connected with the general network. In addition, there are other applications which are not (yet) covered by MAP. These include for example, office applications from the design and process planning areas. Strict realtime demands can also necessitate their own (simpler and hence quicker) communication link. This means that MAP must be connected to other network services or concepts. The parallels between the TOP (Technical and Office Protocol) and MAP standardization gave rise to the comprehensive standard MAP/TOP 3.0 which makes an enterprise-wide network possible (see Fig. C.VIII.09).

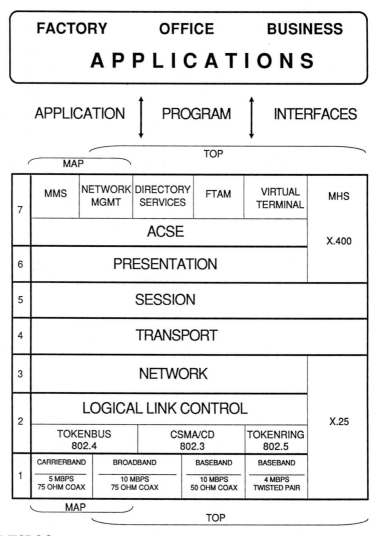

Fig. C.VIII.09: MAP/TOP 3.0
Source: *EMUG*

In the "Computing Network for Manufacturing Applications" (CNMA) promoted by ESPRIT, in which six European manufacturers and four users are working together, MAP/TOP-compatible European products are being developed and tested in industrial installations. Within the MAP architecture the switch-over to other network concepts is explicitly anticipated. Hence MAP becomes a backbone network for further dedicated sub-networks. Fig. C.VIII.10 represents a possible pattern for a total network architecture. GATEWAYS, ROUTERS, BRIDGES and BROADBAND COMPONENTS may also be individually implemented (see *Simon, Kommunikation in der automatisierten Fertigung 1986*).

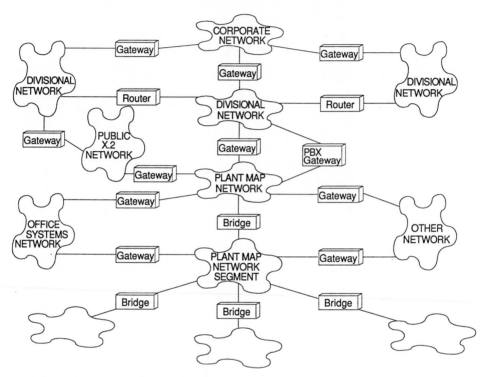

Fig. C.VIII.10: MAP as backbone network
from: *Suppan-Borowka, Simon, MAP in der automatisierten Fertigung 1986, p. 125*

Gateways link networks with differing protocol structures at the higher layers. Here, the so-called cell concept of MAP comes into its own. Cells are independent sub-networks with their own protocol architectures. This makes sense for networks with realtime capabilities, in which the cumbersome administrative work of layers 4 - 7 can be omitted (see Fig. C.VIII.07). A gateway links MAP with those realtime networks, familiar, for example, under the term "Proway", in that the full MAP protocol architecture on one side is lined up against the protocol architecture of the cell on the other side, and they are linked layer by layer via protocol transformation (see Fig. C.VIII.11). This kind of gateway, which allows the link up of diverse network specifications, is also anticipated for other kinds of networks.

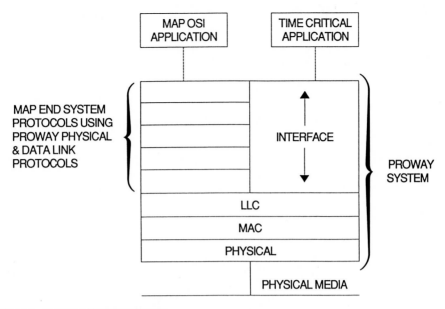

Fig. C.VIII.11: MAP/PROWAY SYSTEM
from: *Janetzky, Schwarz, Das MAP-Projekt 1985*

Routers are of special significance within the network concept, because they represent a relatively efficient linkage. Routers can, if necessary, connect different layers 1 and 2, which can then be linked to layer 3 by means of a unified network protocol. In this way links with public networks, e.g. Datex-P, are possible.

Bridges can join together diverse sub-networks which possess a unified address structure.

Broadband components, such as branches, equalizers and boosters, allow a structuring and extension of the network at the physical layer.

Fig. C.VIII.12 depicts a MAP infrastructure already presented at SYSTEC 88.

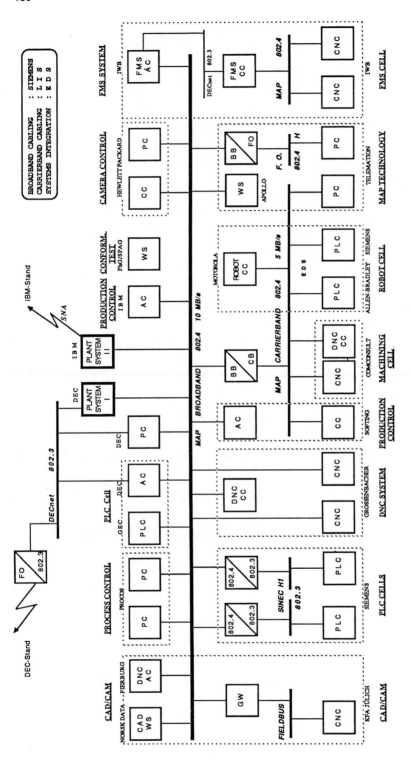

Fig. C.VIII.12: MAP infrastructure

EMUG MAP EXHIBITION SYSTEC 88

3. Data Exchange

In the context of the third layer of Fig. C.VIII.03 the various software systems for CAD/CAM and PPC are connected such that data from one area are transferred to another area via an interface file. The use of interface files is a classic data processing tool for the connection of separate applications systems. Although this kind of processing derives from batch processing, it is nevertheless applicable in the context of interactive processing systems. In such cases individual data records are exchanged via so-called mailbox, message or action systems.

Applications systems must be altered such that the data to be exchanged are generated by the supply system in a form that the receiving system can process; alternatively, special reformatting programs must be produced. This kind of link is, therefore, designed specially for the systems to be connected, and cannot be intended for general purpose use. It presupposes that a standard format for the transfer process can be used, for which pre- and post-processors must be provided. Using this link it is possible to transfer a bill of materials generated by a CAD system to the PPC system. It is, of course, not possible to switch in free form between the systems using query languages. Rather, data exchange is established using precisely defined program functions. This kind of link is currently offered, or being developed, by several CAD/CAM and PPC manufacturers.

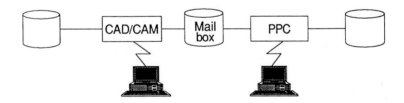

Fig. C.VIII.03,c: Layer 3: Data transfer between systems

4. Common Database

At the fourth layer of Fig. C.VIII.03 a considerably closer system integration is achieved by the systems having access to a common database. Here bills of materials and CAD data, or work schedules and NC data are not held in seperate databases; rather, both application areas use the same database. This can lead to a high layer of data integrity, since updates from one area are immediately available to the other area. This layer of integration presupposes, of course, that a unified data structure is defined for both areas and that a unified database system is installed.

182

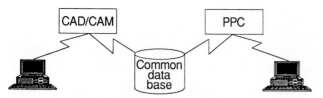

Fig. C.VIII.03,d: Layer 4: Systems sharing a common database

The main arguments for the use of database systems are (see *Dittrich, Datenbankunterstützung für den ingenieurwissenschaftlichen Entwurf 1985*):

- **Support of data integration**: Data stored in the database can be made available for a variety of uses.
- **Application related data structure**: By the definition of sub-schema subsets of the database can be made available to individual applications.
- **Support of consistency**: Database systems provide assistance in the support of data consistency, particularly in the case of deletion and amendment.
- **Support of multiple user systems**: Appropriate system technical measures ensure that competing access to data entries do not endanger the integrity of the database.
- **Data security**: Data are protected against service failures by database system tools.
- **Data independence**: The application and design of databases are largely independent of the physical data storage system.

Alongside these factors, the use of database systems supports applications systems programming. The query languages assigned to the database system allow even inexperienced users to put together flexible evaluations on an ad hoc basis. When we examine the demand that the use of proven database systems, such as are available to the PPC area, also be planned for the geometry- and technically-oriented areas, it must be recognized that the use of classic database systems for so-called non-standard applications can lead to substantial performance problems (see *Hübel, Datenbankorientierter 3-D-Bauteilmodellierer 1985*). For this reason the CAD/CAM area has tended to work up to now using conventional file organizations. In contrast to PPC applications, CAD applications have the following characteristics:

- The data structures are very complex, and only rarely do identical data objects appear. In classic PPC applications, however, the data structures are simpler, and a multiplicity of similar data objects are handled.

- In the course of drawing creation, design variants are produced, which need to be administered in chronological order of creation. In classic PPC applications, however, the chronological order of the amendment history of data records is unimportant.
- Transactions within the design process can last for an extremely long time (e.g. the creation of one drawing can extend over several days). In contrast, a transaction within a PPC system, which takes a database from one consistent condition to another consistent condition, takes place within seconds.
- The creation of geometry data requires attention to multiple, extremely complex consistency requirements, relating to design rules and processing plans. In contrast, the consistency requirements of PPC systems are considerably simpler.

This list of requirements disallows the demand for uncritical application of standard database systems in the geometry area. There are three possible solutions, though: First, standard data models can be extended using CAD data management packages. This possibility has the advantage that it allows simpler access to classic database applications (in particular in PPC). Of course, there are disadvantages involved, affecting performance, for example. A second possibility is the development of new database systems, in which these problems are more adequately handled. This implies the use of dedicated database systems, which can only be linked to existing applications with difficulty. A third solution to this problem might consist in defining a common database core, which can then be extended for different applications using software modules (see *Fischer, Datenbank-Management in CAD/CAM-Systemen - no date - Fig. 3; Dittrich, Datenbankunterstützung für den ingenieurwissenschaftlichen Entwurf 1985, p. 124*). In concrete terms this means, for instance, that bills of materials from the PPC system and the descriptive and classificatory data of the CAD system are held in a common database, while the geometry data are kept in another database. In addition, the database has special functions for CAD data management, which allow extended data integrity safeguards, and the special handling of transactions in CAD applications.

5. Inter-Application Communication

In inter-application communication (see Fig. C.VIII.03) transactions of one system can access transactions of the other system. This represents a real staggering of program functions. Ultimately, this means that operating systems and database systems from different application areas can communicate with each other. Such a high degree of integration cannot be expected in the near future. Nevertheless, initial tendencies in this

184

direction are recognizable at both the computer technical layer (LU 6.2) and in the new developments in integrated applications software.

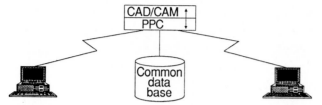

Fig. C.VIII.03,e: Layer 5: Inter-application communication via program integration

c. CIM Data Handlers as Integration Bridges

Since the implementation of a comprehensive inter-application communication for CIM cannot be expected in the short run, the Institut für Wirtschaftsinformatik (IWi) in Saarbrücken is developing a **CIM handler**, which is a central system for controlling information exchange between CAD, PPC and other standard software systems, such as

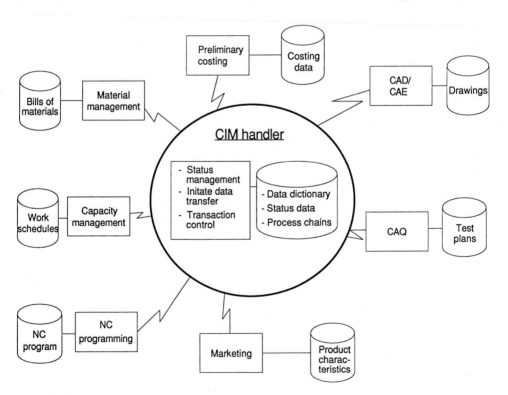

Fig. C.VIII.13: CIM handler concept

CAP, CAQ, and DNC (see Fig. C.VIII.13). It utilizes all the available integration layers and combines them into a global concept. It ensures that changes in one of the systems involved, insofar as they are relevant to the other systems, are transferred either automatically or after further processing, so that the consistency of the operational database is maintained. Manual data transfer between the systems is then no longer necessary.

The interfaces between the standard software components and the CIM handler are so universally specified that several CAD or PPC systems can be alternatively connected up without requiring changes to the CIM handler.

Prerequisites for the implementation of the CIM handler are:

- **Hardware links** with the relevant base functions. In particular, it must be possible for an application to communicate at any time with other computers and systems (terminal emulation, initiation of file transfers, etc.). At the end of the communication, return to the "same place" in the original application must be possible; e.g. it must be possible at any point in the production of a drawing at a CAD workstation to conduct a dialog with another computer, from the financial system, for example. When the interaction is completed production of the drawing can be resumed, without the CAD system having to be started up anew.

- **User exits**: The standard software systems to be connected must provide user exits, so that standard functions can be linked up to in-house produced systems. For example, many CAD systems are already configurable according to customer demand, such that functions can be defined which combine CAD macros (e.g. the function "store drawing") with in-house produced programs ("grant release order for drawing"). The same requirement also applies to the other standard CIM components; for a PPC system this would mean, for example, that for every call up of the function "change bill of materials structure" a program can be activated which collects the altered bill of materials data and passes them on to the CIM handler.

- Visibility and **documentation of the databases**: This is necessary to allow data exchange transactions and free information possibilities between the system components with the help of the CIM handler. The use of (relational) database systems with the corresponding tools (queries, report generators) is a help.

The **functional scope** of the CIM handler is described below:

- **Data exchange transactions**: Complete data transfer possibilities for currently generated data between all incorporated systems are ensured via the implemented data exchange transaction. For each system a transaction menu is provided for the system user which can be produced automatically and which is coordinated with all the connected systems. From this menu the desired data exchange transaction can be selected and started. The initiation of a data transfer transaction is always carried out by an authorized user in the context of alteration clearance. An **example** will clarify the data exchange transaction:

Traditional procedure: A change in a design drawing (e.g. replacement of a component for technical reasons) necessitates changes in the bill of materials and the work schedule for the relevant part; at the moment, these changes are generally communicated to the PPC system in an appropriate (usually written) manual form. The conformity of the design bill of materials and that of the PPC system is not ensured in computer terms, that is, failures can arise because changes are not communicated correctly or in time.

Procedure with CIM handler: When a drawing alteration receives clearance, the CAD system prepares relevant data (including, for example, designer's remarks for other operational departments) for all the other systems (PPC, CAP, CAQ, parts classification system, etc.) and passes them on to the CIM handler (see Fig. C.VIII.14). The handler then carries out plausibility checks on the databases of the systems involved ("Is the new component already defined in PPC?"), splits up the data delivered from the CAD system into alteration data for the individual systems and transfers these (for automatic entry or as raw data for further processing) to the relevant systems. At the same time, the CIM handler automatically applies

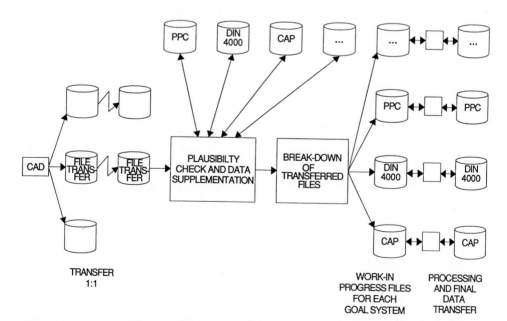

Fig. C.VIII.14: Data exchange transaction procedure
Source: *Gröner, Roth, Konzeption eines CIM-Managers 1986*

status markers, which show how far the processing of the alterations in the individual systems has progressed. When the alteration processing in all systems is completed, the consistency of the operational database is restored.

- **Information transactions**: The system also automatically produces a menu for information transactions which is analogous to the menu for data exchange. The user of a system can fetch files from connected systems by choosing from a menu function using either pre-defined or free-formulated requests. Examples might be database queries about parts information to the PPC system, access to a parts classification system, access to inventory and capacity information.

- **Inter-system status management**: The task of inter-system status management is to monitor the processing stage of all data exchange transactions between the communicating systems using status markers. As a result the stage of each individual transaction can be determined in detail at any time. All status markers are assigned automatically. However, the system manager does have the power to intervene in response to certain system states and to alter the status markers manually.

- **Trigger concept**: The CIM handler makes it possible, using status markers applied by inter-system status management, to initiate or block further transactions, to release or block data records and to carry out plausibility checks of all kinds. In this way it is possible to send data automatically to another operational location after it has been processed in one system, and to initiate further processing there.

- **Data dictionary functions**: The CIM handler contains data dictionary functions, that is, it "knows" which data (data fields, formats) the various connected systems administer, which alterations they carry out, and how these alterations need to be processed by the other systems. On the basis of this "knowledge" it can control the exchange of data between the system components.

- **Customizing and free configurablilty**: The CIM handler allows the user to configure the system freely at initialization of the system. It manages all the connected systems as well as the installation transactions for all the components to be integrated into the total system. Installation transactions check the availability of the components needed for the installation, and record the transactions being initiated or modified by the new system in a transaction index. A transaction database consists of all individual transactions, which can be freely configured to form collective transactions.

This concept developed at the Institut für Wirtschaftsinformatik (IWI) represents a realistic possibility for implementing comprehensive CIM process chains using currently available means (terminal emulators, networking, data dictionaries, relational database systems). It has already been implemented by manufacturers (e.g. Hewlett Packard with the CIM Server system) (see *Hewlett Packard, HP CIM Server 1988*).

IX. Implementation Paths

In general, introducing a complete CIM system in all its breadth would overstrain most enterprises, so that a step-by-step procedure is required. The discussion of CIM sub-chains considerably reduces the complexity of the CIM principle.

Various implementation paths for implementing a complete CIM model can be followed depending on the priorities of the particular case.

Fig. C.IX.01 represents typical introduction sequences depending on different starting positions. Each implementation path leads to the same global CIM implementation. In this way individual starting positions and current problems can be taken into account, without jeopardizing the end result.

For example, if the current problem is one of meeting a customer's quality assurance requirements (CAQ) the existing production planning and control system can be linked with CAQ by adopting test plans (see the lower, boldly outlined path in Fig. C.IX.01). Thereafter, this can be linked with an operational data collection system for feedback of quality data. CAD is then introduced in a subsequent step.

The current problem for another enterprise might be that of linking CAD with NC programming, so as to shorten the delivery time on variant orders. The next step would then be to introduce DNC and thereafter to reorganize production planning (see the upper, boldly outlined path).

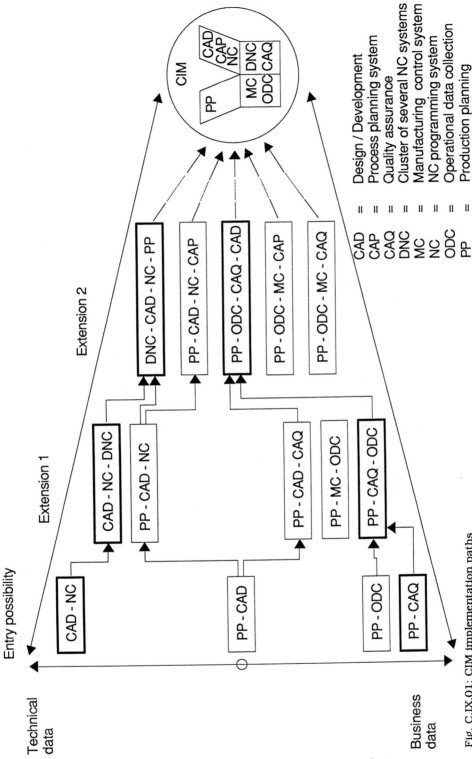

Fig. C.IX.01: CIM implementation paths

D. CIM Implementations

Having developed a procedural approach to introducing CIM, several CIM implementations will be presented.

Since the introduction of a complete CIM solution requires several years, none of the examples presented can be regarded as a complete "CIM enterprise". Nevertheless, in each case there is a recognizable relationship to a general CIM structural plan which is gradually filled out by the implementation of sub-systems. As a result, the examples demonstrate differing areas of emphasis in the implementation sequence.

Earlier editions of this book presented CIM prototypes which had been created for demonstration purposes in various manufacturers' CIM centers. They have been eliminated from this edition because these prototypes have already become familiar to many interested parties from personal experience, and in the meantime more "authentic" CIM implementations are becoming available which are more informative for the reader interested in implementation.

The six examples from the German industry, which are described in relative detail, are contrasted with five American CIM implementations. They should demonstrate on the one hand, that the idea of CIM has in the meantime come to be viewed in the USA, as in Germany, as the comprehensive task of the computer steered firm. Furthermore, the examples of the products of important CIM suppliers (IBM, DEC and HP) demonstrate the experiences of these manufacturers, which are also incorporated in the development of CIM products for their customers.

I. CIM Solutions in the German Industry

a. CIM Implementation at *ABS Pumpen AG*, Lohmar

(Dipl.-Ing. (TU) Klaus Blum, ABS Pumpen AG, Lohmar;
Dipl.-Kfm. Wilfried Emmerich, IDS Prof. Scheer GmbH, Saarbrücken)

ABS Pumpen AG, with its headquarters in Cologne, is one of the leading manufacturers of submersible motor pumps for sewage and waste water drainage. The pumps cover a performance spectrum from 0,2 kW to over 500 kW with an extraction rate of up to 11,000 litres/sec. The company group employs a total of 1,300 employees in 5 production locations

(Germany, Ireland, France, Brazil and USA) and includes numerous international marketing companies. The enterprise, which was founded by Albert Blum in 1959, achieved a turnover in 1989 of DM 200 million with a high proportion of exports.

The principal consumers of ABS products are the plumbing supplies industry, and in the context of increasing environmental investment, a rising number of municipal and industrial bodies both in Germany and abroad.

In addition to the usual cutting procedures in production ABS uses a variety of modern techniques and modern materials in the processing of plastics and stainless steel.

In order to handle the ever-increasing turnover it is necessary to accelerate the internal organizational processes. The use of modern information systems is essential. In July 1988 IDS Prof. Scheer GmbH was commissioned to develop an outline concept for information processing. The overriding aim of this concept was to be the reduction of the administrative component in the order throughput time.

At that time information processing was characterized by the use of heterogeneous hardware systems and a multiplicity of applications software systems. These systems had been delivered by various manufacturers and specially adapted to user-specific requirements. Given the substantial modifications, the standard software package "release change" could not be effected. The applications systems for the business and logistic

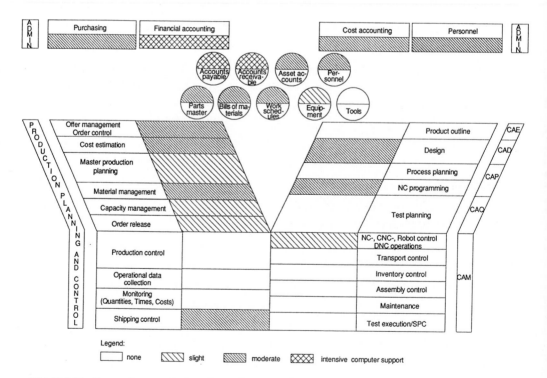

Fig. D.I.01: Computer penetration (initial situation)

functions (purchasing, material management, sales, production planning, cost accounting, financial accounting, pay-roll accounting) were linked together via batch interfaces. The daily data exchange meant that the systems were operating on data of varying up-to-dateness.

The computer penetration in the initial situation is shown in Fig. D.I.01.

The appraisal of the current organizational processes in the specialist departments was documented using process chain diagrams. The weak points that this revealed were essentially a result of the non-integration of logistic CIM functions.

The following framework conditions were formulated for the development of the planned concept:
- reduction of system diversity (hardware and software),
- more extensive use of unaltered standard applications software,
- applications integration.

The demands on the future information system were established in several working parties. The working parties consisted of employees from the specialist departments and the organization department of ABS Pumpen AG and the project team from IDS Prof. Scheer GmbH.

The PPC system had to cover the customer-order-oriented one-off production, primarily of large pumps, in the main works at Scheiderhöhe through to serial production of components and end products at the Wexford works in Ireland.

Pump design requires a 3D CAD system which also allows the definition of free-form surfaces. The integration of NC programming is also required.

For quality assurance, tests on goods received, production and end products as well as test equipment management and planning are necessary functions of a CAQ system. Product liability requirements as regards quality documentation can be fulfilled by an integration of the PPC system (order and product data) and the CAQ system (quality certificates).

The choice of software was made in the following stages:
- evaluation of the completed suppliers' requirements catalogues,
- presentation of the system by the suppliers using test data from ABS,
- visits from user referees.

The decision was made to use the PIUSS-O system from PSI for the PPC system, BRAVO 3 from Schlumberger for the CAD system, and QUALIS from Softsys for the CAQ system.

The hardware infrastructure was completely changed from the existing system (IBM/38 and HP 900/200) to DEC-VAX computers and workstations. For the other applications systems the availability of these types of computer is absolutely essential.

With these decisions, the essential prerequisites for the integration of the technical and organization-planning applications systems were achieved. The further integration will also be implemented to include distant sites (starting with the works in Ireland).

The introduction of this system began in the second quarter of 1989. The PPC system was put into productive use at the beginning of 1990. The financial and cost accounting modules of the VAX-ProFi system as well as the personnel information system PAISY was introduced at the end of 1989.

The CAD and CAQ systems will be introduced gradually during 1990.

After introduction of the PPC system the system for job-shop control and operational data collection will be chosen and implemented.

The following project organization was established for the entire CIM project:

The CIM project group with the project leader and two further employees report to the management board and the directors. The project leader is employed full-time with the handling of the project, whereas the other employees from the CIM specialist areas are available on a part-time basis.

The CIM project group coordinates the working groups for the introduction of the CIM components PPC, CAD, CAQ, job-shop control, operational data collection, NC, cost accounting, financial accounting, pay-roll accounting as well as the specification of the operational structure, the computing infrastructure and the introduction of the office communication system All in One.

The leaders of the working groups are supervisors or employees from the relevant departments. Each working group leader is supported by a member of the CIM project group.

For the introduction of the PPC system the working group is broken down into working parties which are responsible for the material management, sales, customer services, primary data structures and production areas.

On completion of the plan described the penetration represented in Fig. D.I.02 will be achieved.

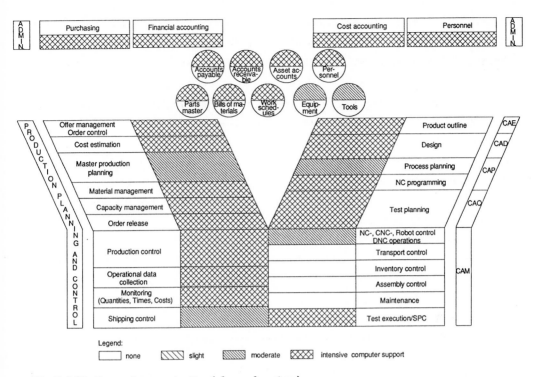

Fig. D.I.02: Computer penetration (planned system)

b. Information and Control System in the Production Area as a Central Agency for the Dataflow of a CIM concept at BMW AG, Dingolfing

(Dipl.-Ing. (FH) Richard Baumgartner, BMW AG, Dingolfing)

With the increasing degree of automation it is becoming both more difficult and more important to comprehend the nature of manufacturing installations and processes. The computer-technical information available is becoming more copious, decision times shorter. Hence automatic preparation, the production of protocols and the archiving of available data is absolutely essential to the comprehensive documentation of the production process. Intelligent, autonomous, stored programmable controls possess a multitude of possibilities for data capture and supply. This covers production and failure data from several hierarchy levels in diverse data formats. These data are important both for the functional-organizational unit concerned, and as input for upstream or neighbouring systems.

This information flow, therefore, represents an important prerequisite for integrated systems, such as CIM. In general, difficulties arise only in linking up the diverse technical systems and creating defined interfaces. Recognizing this, the Bayerische Motoren Werke

AG has gradually introduced information and control systems in several production areas of its plant in Dingolfing. From the pilot projects it was possible to obtain important information relating to technical performance, data recording and processing, employee acceptance and efficiency. The first applications were installed in production equipment that was already in productive use. This procedure certainly implies considerably higher technical costs of linking and integration, as regards efficiency considerations and system effectiveness, however, it is feasible to build up on an existing base.

The aims of the implementation are:

-   to deliver current information relating to production data and disturbances on a continuous basis (also to higher level systems),
-   to record manufacturing events and stoppages and convert them into appropriate data forms.

This information should serve to:

-   create a basis for organizational decision making,
-   monitor highly automated production lines and facilities,
-   allow optimal capacity use and scheduling,
-   analyse weak points,
-   monitor the effects of optimization, by carrying out simulations for subsequent stages,
-   allow the controlled planning of preventative maintenance,
-   obtain consolidated key planning data concerning actual running times, manufacturing buffers, degree of chaining, and chaining systems,
-   create modules at the machine level for moving in the direction of computer integrated manufacturing.

The approximation of operational processes to a complex CIM system can only be undertaken in successive steps. This stepwise implementation must nevertheless be based on a planned global concept, which is rigorously followed in the implementation of the individual steps. Necessary amendments to user-specific details should not be ruled out. These will consist of organizational adjustments as well as the creation of the necessary technical requirements for communication between the various systems in use. An information and control system thus represents an initial computer-technical linking of the planning, control and management levels with the execution levels (see Fig. D.I.03). The data flow runs from the planning and control levels to the execution level, and feedback from the execution level flows back to the higher levels systems.

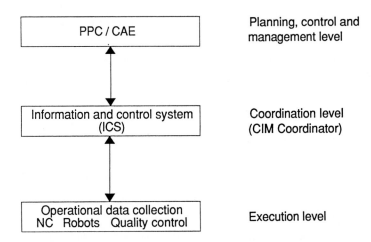

Fig. D.I.03: Information and control system as the connecting link between the planning and control level and the execution level within a CIM concept

In practice the PPC and CAD systems are installed on HOST hardware, whereas at the execution level it is mostly smaller autonomous systems that are in use. The control system, therefore, links up various levels of operational processes using fundamentally different hardware (see Fig. D.I.04).

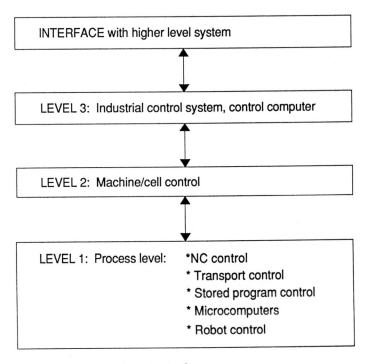

Fig. D.I.04: Information and control system levels and interfaces

The tasks of the CIM control centers introduced at BMW are primarily evaluation, communication, planning and control functions in the areas of equipment, job-shop and maintenance planning. They provide information to the higher levels about the operational state (the status of plant, machines and orders) in usable form, serve as initiative or input for new planning processes, for the control of inventory procedures, or the internal analysis and evaluation of the organizational/functional areas.

The control system must display the raw data from the autonomous systems, prepare it, make it available and archive it. To make available means, in this context, to condense the data in various aggregation levels with diverse cycles and to adjust the data to the various systems.

The manufacturing systems under consideration consist of independent production areas with diverse production facilities (processing and welding machines, as well as conveyance and storage systems). These are controlled by autarchic stored program controls.

At the process level (see Fig. D.I.04) disturbance, production and quality data are collected, these are prepared in protocols and statistics at shift change, then condensed, passed on and archived. The evaluations relate to pre-defined time scales depending on the level of aggregation (day, week, month, year) and provide a complete picture of the production process.

Through continuous, current information relating to
- finished quantities,
- number of parts in store,
- quality results,
- number of breakdowns,
- current down time,
- aggregate repair and down times,

production control, maintenance and the higher level systems always have access to the most important data needed for sound decision making.

Once it was established within BMW that an information and control system should be introduced as a pilot project, an operationally experienced and technically qualified project group produced a concept as a basis for discussion. Details of executions such as system handling, message texts, message types, time considerations, etc. were determined along with the affected locations within BMW. The bulk of the coordination was restricted, however, to the two main system users: production and maintenance. The works council were also informed at this time.

In addition to guidance in the use and operation of the system, manufacturing and maintenance personnel were given extensive local instruction as to the totality of the interdependence and the objectives. As a result of this open treatment of the issue the control system was implemented and accepted without problem. Subsequent projects were conducted analogously to the pilot project.

Experience has shown that, the more use is made of the data generated, the more effective and efficient the control systems can be.

Clearly specified maintenance and shut-down work together with punctual optimizations can significantly reduce the down-time for production installations. Parts production is increased and stabilized, and the number of breakdowns reduced. For production lines unequivocal testimony as to the productive running time, machine performance and capacity with respect to the individual machines is provided, which influences the higher level planning and control systems.

It has also become clear, though, that this kind of project can only be assessed up to a point using classical efficiency rules. It is necessary to define the value to be placed on the following factors, and how they might be included in economic efficiency calculations:

- production security,
- current knowledge of capacity reserves and production bottle-necks,
- the value of sound experience in new planning,
- verifiable quality in the operation of production installations,
- verifiable product quality,
- current evaluation of production problems,
- quality and up-to-dateness of planning.

The use of information and control systems at BMW has shown that, in structuring highly automated production installations, systems for production monitoring, production control, maintenance, maintenance planning and production planning are economically justifiable and, in the long run, urgently needed. However, extensive technical details need to be resolved in the implementation and introduction. Efficient use of the system requires not only the qualification of the employees, but also acceptance of the system by those who operate it. In the meantime, BMW generally installs control systems along with the creation of new production installations. Interface problems between the various sub-systems are then more easily solved. Furthermore, more cost-optimal structures can be achieved.

c. Computerized Flexible Production Chains for Pressed Sheet Metal Parts at MBB, Bremen

(Dr.-Ing. Ulrich Grupe)[1]

The market-related fluctuations in production rates on the current airplane construction program, along with the resulting need to reduce costs, has necessitated the use of flexible production systems for the multitude of diverse sheet metal parts, such as have been used successfully for some time for milling and drilling operations. Modern production units begin to display their full productivity when they are electronically (usually numerically) controlled, linked up with suitable peripherals and led by effective information processing. Hence a computerized automated production system was developed, referred to in short by the acronym CIAM (Computer Integrated Automated Manufacturing). This fundamental concept led to the achievement of CIAM forming at MBB in Bremen.

Pressed sheet metal parts are generally produced in four essential work steps. First of all, the platinum outline is produced in flat form, whereby the edge requirements for airplane construction favour outline milling over other cutting techniques (laser or pressurized water cutting). Thereafter heating ensures that the material is easily formable and adds to the final stability of the part. Shaping is carried out by pressing the metal plate against a forming tool using a flexible upper tool in the form of a rubber cushion or a rubber membrane. Usually, the part has achieved its final form at the end of this process, and fulfils the conditions prescribed by design, such that, after surface protection, it is ready for assembly. Only where the geometry of reshaping poses technical problems a manual reworking of the form is required.

The planning phase investigations relating to storage, conveyance and supply areas between the individual stations were based on a detailed analysis of the parts spectrum and material flows. This showed that an installation of the new production facilities in available premises would have required too many compromises. Within a new production space all the sheet metal manufacturing functions from material receipt to the finished part could be integrated, and the structure of the plant could ensure an optimal material flow, as is shown in Fig. D.I.05. A few essential aspects of the more important manufacturing facilities are explained below:

---

[1]   I would like to thank Dr. Humbert, ex-head of Department TF4 (Data Processing Systems for Production) at MBB (Hamburg) and now manager of the German Airbus GmbH (Hamburg), for revising this article.

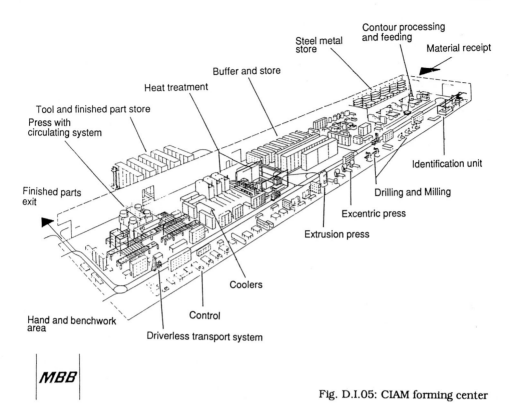

Contour processing and feeding

Steel metal store

Material receipt

Buffer and store

Heat treatment

Tool and finished part store
Press with circulating system

Identification unit

Finished parts exit

Drilling and Milling

Excentric press

Extrusion press

Coolers

Hand and benchwork area

Control

Driverless transport system

MBB

Fig. D.I.05: CIAM forming center

Contour milling and pre-drilling of fixing and rivet holes is carried out on a numerically controlled drilling/milling center. This can take up to 12mm thick lots of standard metal sheet (1.2m x 2.5m), stack, bore, drill and separate them. To make efficient use of this facility it is necessary to select suitable parts, to position them on the standard sheets in material optimizing lots and to recognize and fix or discard the offcuts resulting from overlapping cutting paths. For heat treatment a special annealing furnace is installed, along with the normal equipment for other heat processes. The control-technical concatenation of several transport components ensures that the incandescent material is cooled off in a delay-minimizing mixture of glycol and water within 7 seconds of incandescent temperatures, then rinsed, dried and returned. A special press allows a maximum press load of up to 700 bar for maximal palette sizes, which can be increased up to 2500 bar by reducing the palette size. The short press cycle of one-and-a-half minutes is exploited by disconnecting the supply and removal channels by means of a palette circulation system. The main processes described, along with the other facilities depicted in Fig. D.I.05 are connected by a driverless transport system. On an inductive track of about one kilometre four vehicles serve 35 handling stations, as well as commissioner, intermediate and buffer stores, whereby euro-norm palettes can be delivered or removed from up to four stories using hoist devices.

202

The information processing which accompanies the production process and allows efficient control and the embedding of this manufacturing area in the entire productive environment is split up into several hierarchically structured levels.

At the highest level of this structure system in use throughout the enterprise are implemented, including both those used for the technical functions of machine design and NC programming and those for the administrative functions of work schedule data management, material requirements estimation, and production order release, so that uniformity of techniques and of planning data are ensured for the interaction of the manufacturing areas. The hierarchy of computer functions for CIAM forming shown in Fig. D.I.06 represents these tasks defined as the central control level simply as a superior communication-linked level. Fig. D.I.06 represents the specific computers used exclusively for the production chain and their functions.

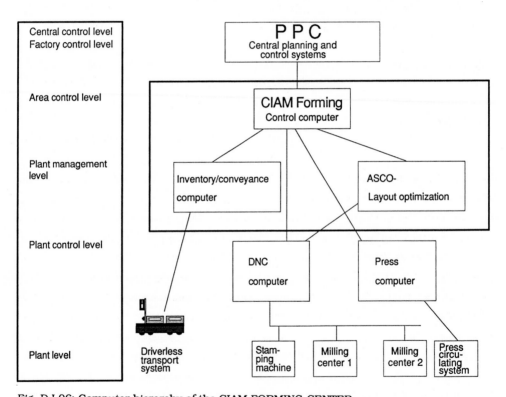

Fig. D.I.06: Computer hierarchy of the CIAM FORMING CENTER

At the highest specific layer, denoted the area control layer in Fig. D.I.06, can be found a summary of the scheduling and coordination activities in the form of the production control computer. This is closely linked with the plant management layer. Here hand- and benchwork commissions, batches for heat treatment, and "menus" for re-forming on a press palette are arranged, and a transport plan is produced for part batches from received production orders. Furthermore, order-related flexible nesting of sheet metal outlines in the standard sheet also belongs to this functional level, where the actual nesting algorithm is provided by the ASCO (Automatische Schnittplan-Optimierung - automatic cutting plan optimization) system from the firm Krupp. The material flow computer schedules and monitors the driverless transport vehicles and manages the entire stores, as well as the buffering and commissioning between the manufacturing steps and the providing of individual parts for assembly and the necessary equipment within CIAM forming.

At the next level down the acquired and released instructions for parts manufacture are transformed within a realtime, time-sharing system into operative controls. The numerically controlled machines receive their control data directly via a loop network. The press computer monitors the palette circulation and controls the relevant press parameters for each palette. The system layout is such that the material flow and production control computers are considered as congruent systems, so that realtime functions of material flow control make reserve capacity available in case of disturbance, to allow a quick restart.

Beneath the realtime system levels are the individual machine controls, which are normally considered as an integral part of the production stations. Only in the case of the conveyance system an additional layer is added in the form of a central control above the on-board vehicle computers and routing control.

d. CIM Implementation at Metabowerken, Nürtingen

(Dipl.-Ing. Manfred Heubach, Metabowerke, Nürtingen; Dipl.-Kfm. Helmut Kruppke, IDS Prof. Scheer GmbH, Saarbrücken)

1. Developing a CIM Strategy

Towards the end of 1986 Metabowerke[2] began to consider the creation of a CIM implementation strategy for the functional areas:

- production planning and control and operational data collection,
- product development (design, process planning),
- product manufacture (NC programming, quality assurance).

---

2   Henceforth referred to as Metabo.

These considerations gave rise to a CIM strategy which was developed along with IDS Prof. Scheer GmbH as external consultants.

The CIM strategy was oriented towards the following Metabo objectives:
- increased flexibility
  - -- with respect to customers
  - -- in production by the support of various control procedures
- reduction of intermediate stocks
- reduction of throughput times
  - -- in product manufacture
  - -- in product development
- improved availability
  - -- 24-hour operation
  - -- short response times
- improved provision of information
  - -- with respect to customers/marketing
  - -- for central planning
  - -- for management
- future efficiency
  - -- allowing for open computer standards
  - -- high availability
  - -- high ability to integrate with the process level
  - -- easy extendability.

Metabo's CIM strategy consists of the following complexes:
- the formation of focal areas for the application of CIM on the basis of prior actual analyses,
- the development of planned process chains for the functional areas under consideration taking all other relevant requirements for integration with other functional areas into account, including the administrative functions of financial accounting, cost accounting, pay-roll accounting and purchasing,
- allocation of functions to the logical levels (enterprise level, factory level, area level, etc.),
- choice of applications software for the functional areas taking the integration requirements into account,
- creation of a hardware and network strategy,
- creation of a database strategy,
- analysis of economic efficiency,

- creation of a plan for the sequence of introduction taking account of:
  -- logical dependences between the functional areas,
  -- economic considerations,
  -- capacities (personnel, data processing requirements).

Fig. D.I.07 indicates the areas addressed in the context of the Metabo CIM strategy.

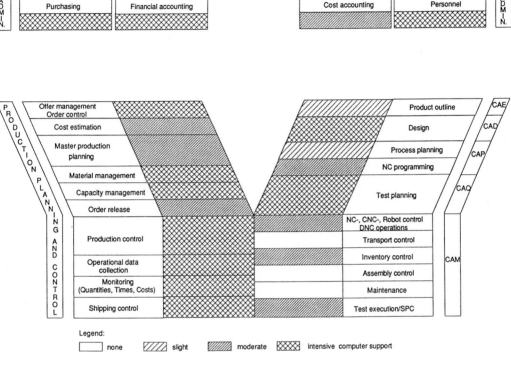

Fig. D.I.07: Computer penetration (planned system)

2. Focal Areas of the Metabo CIM Strategy

The central components of the Metabo CIM strategy included the following long term objectives, which will be discussed further below:

- maximum flexibility and availability via computer application in the production planning and control area as well as optimal integration of process-related systems accompanying production,
- high integration of CAX systems for product development and manufacture and their incorporation in the immediate production control functions,

- optimal support of integration principles and improved efficiency for computer operation by means of a suitable hardware and database strategy.

2.1 Levels Approach in the Production Planning and Control Area

A distribution of the functions of the production planning and control area to different levels was carried out.

The functions of master planning, material and capacity management are assigned to the **planning level**, the production control and data collection control functions, however, are located at the immediate **control level**. Order release from the higher level PPC system, therefore, constitutes the interface between the planning and control levels.

This also implies that the subordinate control level only processes a subset of the total stock of PPC system orders. This also applies to the management of master data (workplace information, personnel data, shift models for personnel and machines), whose content and scope are abridged to those relevant to production control. The reconciliation of master and transaction data is effected by computer either in online processes or as a batch job. Online transfer of PPC orders to the control level is necessary, for instance, in the case of order cancellation and quantity or deadline changes, assuming these orders had already been passed on to the control level. Another example is that of rush orders that have to be transferred immediately to the control level.

In addition, the control level incorporates management functions for resources relevant to production, such as materials, NC programs and equipment. This resource management is reconciled by machine with the higher level planning levels as regards stocks and stock movements and again contains only the relevant parts of the information files held at the planning level. Recording of stock movements is effected immediately at the control level, the movements are then passed on by machine to the higher planning level. An analogous procedure is adopted to handle equipment movements.

The third level, **the process level**, constitutes those computerized functions which either bring order-related data directly to the workplace, or provide feedback directly from the workplace.

This includes machine data collection functions which directly inform the control level of breakdowns, for example. Other functions of this level are inventory control and conveyance systems, which receive stock addition, withdrawal and transport orders from the control level and initiate the corresponding material movements coordinated with production.

Administrative systems

PLANNING LEVEL

Wages and salaries	Cost accounting		Purchasing
Fixed asset accounting	Financial accounting		Inventory management

Common primary data management

Jobs | Prodn. plan | Parts | BOMs | Work schedules | Equipment | Stocks | Test plans

Planning

Customer order handling/Master production scheduling	Demand determination		Detailed planning/ Order release	- Material availability
Rough capacity planning	Medium-term capacity planning	Production orders		- Simulation

CONTROL LEVEL

Control/monitoring   **Dedicated management systems**

Production control	Production orders	Shift model	Tool management	NC program management
Operational data collection	work-places	Personnel	Inventory control	CAQ Test order management

OPERATIVE LEVEL

Manual jobs	Racking storage	CNC operation	CAQ operation

Op. data collection/ local controls

Fig. D.I.08: Levels approach in the production planning and control area

Fig. D.I.08 shows a simplified form of the Metabo levels approach in the production planning and control area.

The **advantages of the levels approach** can be summarized as follows:
- **high failsafe** production control by the use of dedicated hardware systems,
- improved **availability** of production-related applications by the use of decentralized hardware; independence from the current availability of the higher level planning computer,
- improved **flexibility** for diverse optimization procedures in the control area (production islands, job-shop production, formation of kiln lots, cutting optimization, etc.),
- improved **communication capabilities** for linking process-related automation systems using decentralized hardware models,
- **future efficiency** as a result of flexibility and communication capabilities,
- separation of the planning and control levels provides better support for the **diversity of functions**,
  -- computer-intensive planning functions and the processing of large volumes of data are assigned to the host computer,
  -- current applications which make high demands on graphics and response time behaviour are transferred to the decentralized hardware environment,
  -- the functions of the planning levels are required during office hours, the control level functions relate to 2- or 3-shift operation,
  -- the user interface in the production control area can be better tailored to the requirements of current control and monitoring (color, graphics, online warning systems, the use of the mouse to avoid keyboard input as far as possible, etc.).

2.2 Applications Integration in the Technical Areas

The CIM implementation in the technical systems of CAD, NC programming and CAQ consists of the following implementation steps:
- management of product-related and resource-related technical data in a unified database, the Engineering Database (EDB),
- integration of the CAX systems with the PPC functions via
  -- linking of design and process planning with production planning and material management,
  -- linking of the decentralized quality assurance system with the corresponding test planning module in the PPC system by means of a unified database in the test plan, supplier and parts master file management areas,

-- linking of the technical functions with the decentralized production control system for the online preparation of geometry information or the order-related current generation of NC programs (long term).

2.3 Hardware and Database Model

In the course of developing the CIM strategy it was decided to orient the technical applications systems (CAD, NC programming, CAQ) and the control level functions (control center functions, data collection control functions) towards open standards.

The UNIX systems were favored in this context for the possibilities they allow in developing a concentration of know-how at Metabo with respect to operating and maintenance tasks, and in providing greater freedom in the choice of hardware and securing the long term effectiveness of software investments.

In this context there also arose the demand for unified system software for controlling interactive processes in the form of a uniform user interface using window techniques. This generates the possibility of setting up CIM-suited multi-functional workplaces with online access to all currently required information from neighboring areas.

To provide optimal support for the communication necessary between the technical areas and the production control areas (combining technical and order-related data needed for the production processes), the choice of appropriate applications systems was made on the basis of the possibilities for linking them with a global relational database system using SQL.

In the choice of applications systems for the Metabo implementation strategy it was, and continues to be, necessary to ensure that the above-mentioned criteria for operating system, window techniques and independent database concept have been implemented by, or will subsequently become available from, the software supplier.

3. Status and Development of the CIM Implementation at Metabo

Once a detailed timetable and action plan had been developed for the various implementation areas, in 1988 Metabo began to implement their CIM strategy. A large proportion of the functions affected by the CIM strategy were already installed, other functions were to be tackled in the course of the implementation timetable. At the forefront of the approach was the assumption that **standard software** would be employed wherever suitable and economically efficient.

In general, it can be said that the following aspects are decisive for the successful execution of the CIM implementation measures:

-   **Considering long term objectives while implementing effective intermediate solutions:**

    The pursuit of strategic objectives in the areas of database application, functional integration and hardware use may, in certain circumstances, not be fully achievable in the short term (e.g. lack of database support). In such cases it is generally worth considering carefully whether expensive adaptations or in-house developments are economic, or whether some intermediate solution might be preferable until the software supplier can support the Metabo strategy in his standard system.

-   **Supporting the CIM concept by generating its acceptance and creating CIM-suitable organizational forms:**

    The installation of new computer systems in the CIM integration context was accompanied at Metabo by personnel measures to create CIM-awareness and the willingness and ability to support the new system. This applies to practically all levels of the enterprise. Training measures and the generation of confidence, as well as the understanding and acceptance of new processes and activity profiles have been of decisive importance for Metabo in achieving its CIM-related objectives.

e. Fully Automated Material Flow and Information Systems in a Siemens AG Factory

(Dipl. rer. pol. Erich Berner, Siemens AG, Poing;
Dr. Günter Friedrich, Siemens AG, Munich)

The firm Siemens AG appears on both sides of the CIM market: as producer in the hardware and software sector, and as user in its own factories and as planning partner with its customers.

CIM is viewed as the integrating element for all areas of technical and business tasks within an industrial enterprise. This applies not only to the automated factory, but also to the concrete aim of using communication and information links to improve and ensure competitive capabilities. The primary aim of such a system is to fulfil customer orders as quickly as possible without having to keep large stocks of materials and finished goods, i.e., to reduce throughput times accross the board from the initial idea to the delivery to the customer.

1. New Directions in CIM Integration

Whereas in recent years primary interest has been directed at the efforts towards integration between the **CAD** and **CAM** areas - that is, the vertical integration between development/design and production planning up to data transfer to the DNC controlled processing machines, Siemens' equipment factory at Poing near Munich has incorporated a further integration area.

The starting point for the CIM model was the recognition that the long term goals of the production area, that is, "reducing throughput times" and "reducing stocks", are not primarily dependent on close links between development/design and production, but on the efficiency of the **material flow**. As soon as a product has attained the technical maturity to be passed on to serial production, the rationalization efforts must be directed at achieving a "continuity" of material and production processes, even for small series. Attainment of this goal requires close synchrony between the predominantly order-controlled systems of the PPC environment and the movement-oriented mechanical systems of automated conveyance and supply (stockkeeping) in the CAM area.

The **integrational objective** was therefore to achieve the closest possible links between the **PPC** system (for **material management** and **production control**) and the **CAM** systems for **conveyance** and **stockkeeping/supplies**. The integration of these sub-systems led to the implementation of an **automated logistic system** (see Fig. D.I.09).

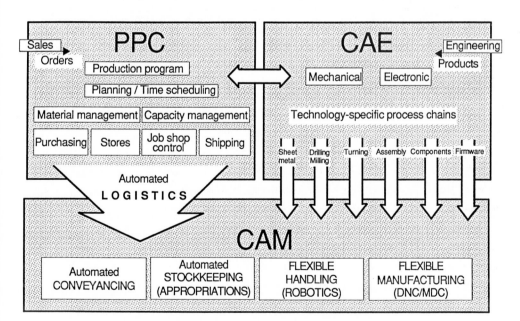

Fig. D.I.09: CIM integration fields: Directions of integration impact

2. Structure of the Production and Material Flow in the Factory

Production layout should already take account of the **flow principle** as far as possible. Sequential ordering of the various functional areas from **receipt of goods** through the mechanical **parts and electronics production** up to **aggregate** and **final assembly**, as is represented in Fig. D.I.10, is certainly of benefit in achieving the logistic goal of shortened throughput times.

SIEMENS

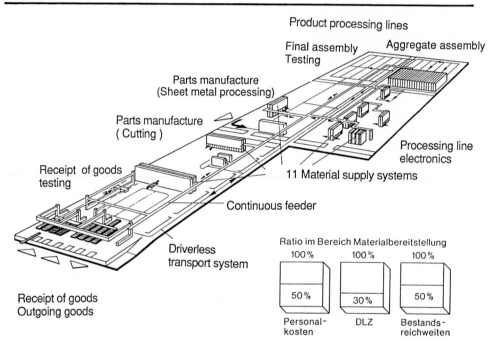

Fig. D.I.10: Material flow concept in the production area

Material flow is closely dependent on the production flow; it starts on several paths from receipt of goods and the associated receipt checks, then proceeds along a **central transport channel** between the production areas leading to the assembly area.

The dominant direction is forwards towards assembly. Branchings, in the form of switch lines which leave the central channel, carry supplies to and from the individual production areas.

To ensure constant availability of materials and parts each of the production areas is assigned a special **material supply system** (MSS) which replaces the former stores.

In the figure these material supply systems can be seen as vertical "slabs" positioned above the production areas.

2.1 Automated Transport System

Two complementary transport systems handle the entire material flow (with only minor exceptions):

1. Continuous feeder
2. Driverless transport system

The "fork-lift truck driver" is therefore a thing of the past.

To have the ambition to transport all parts automatically necessitates the geometric measurement of the entire parts spectrum and its assignment to transport units.

Choice and determination of transport units and their loads:

In the entire factory 30,000 parts had to be measured and weighed.

European standard containers should be used wherever possible as the transport medium. For small to medium sized parts so-called Eurofix containers with a standardized base size but various heights were chosen, and it should be mentioned that a special small parts container had to be developed for many small parts. For larger parts grid boxes and flat Euro-palettes were specified as the transport units and carried by the fixed transport line driverless transport system.

These two automated transport systems carry 30,000 parts/aggregates of various sizes up to finished equipment within the factory.

2.2 Receipt of Goods

Entry into the automated material flow occurs on receipt of goods.

There all deliveries which have not already been packed in the prescribed transport units by the suppliers are re-packed in the parts-specific container types, whereby loads must be tailored to consumption needs in use.

The re-packing instructions are shown on the screen for each incoming order at the receipt of goods point. All further information needed for the subsequent automated transport is automatically supplied by the integrated information system, so that once the goods are passed on to the feed system they reach their ultimate destination without the need for any further manual intervention.

2.3 Parts Production and Flat Assembly Unit Production

Next to the receipt of goods zone comes **parts production**.

Flexible production cells and flexible production systems allow a largely complete processing of several operations, without set-up times having to affect throughput times.

The essential requirement for an uninterrupted production process is the **synchronized supply of materials, tools, devices** and NC programs. The materials and additional equipment required are automatically placed in the stock removal container defined by the material supply system assigned to each production area. This principle applies both to mechanical parts productions (e.g. flexible production systems for drilling/milling or flexible production systems for sheet metal processing) and to electronics (e.g. automatic assembly lines).

In order that the 30,000 parts of different sizes automatically reach their respective destinations without manual intervention, the sender (source) and receiver (destination) addresses of the parts are read from the work schedules of the primary data files and assigned to the machine transport orders. For the employees in receipt of goods and the production areas the only manual data entry required for the automated transport is to inform the system of the individual container numbers and the parts numbers that have been placed in the container, where necessary indicating the quantity. Everything else is then fully automatic.

Container Information System:

In this manner up to 200,000 different containers and about 3,000 palettes/grid boxes circulate through the production areas. The transport system is aware at all times of their contents and current location or destination. A **container-specific transport and information system** ("BIBER") ensures the absolute transparency of the material flow.

2.4 Material Supply System for Assembly

The **assembly** area is the focal point for output creation and here the two largest material supply systems are installed. The MSS initially takes over short term intermediate storage functions for parts needed in assembly. In this sense the MSS is a further development of the already familiar automated racking systems.

Deposition occurs automatically by taking over containers that have been delivered by the two transport systems (continuous feeder and driverless transport system) from receipt of goods or the production area.

The "novel" aspect relates to withdrawals. Here the automated withdrawal functions and subsequent transport of the withdrawn parts are consolidated into a single new functional entity, the **material supply system** (MSS).

The majority of withdrawals (90%) occur completely automatically without manual intervention. This applies to so-called "process parts" for which container types and loading quantities have been established and which therefore no longer need consignment orders.

These process parts are largely removed from stock during the third "unmanned" shift. The containers to be withdrawn are brought by the rack operating devices to the outer area, where so-called "assembly containers" are docked in the storage alleys. These assembly containers ("MOCON") are automatically transported by the driverless transport system from the various assembly workplaces to the pre-determined storage alley whenever the assembly workers request supplies from the MSS.

The principle of this fully automated material supply will be illustrated briefly with the example of **aggregate assembly**. In this area aggregates are assembles in so-called "assembly islands". The island is a rectangular area, which is surrounded by the materials required there, either in assembly containers or in grid boxes or palettes. The employees' workplaces are inside the island (see Fig. D.I.11).

SIEMENS

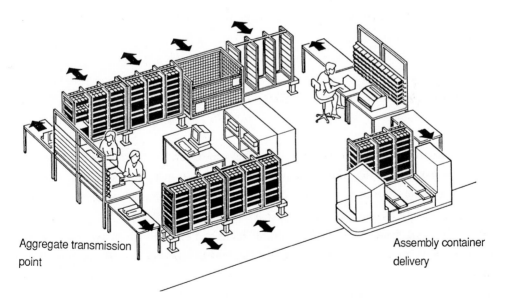

Aggregate transmission point

Assembly container delivery

Fig. D.I.11: Supply principle for assembly islands

The employees remove the parts from the assembly container compartments or from the grid boxes or palettes at the workplace in accordance with the assembly sequence. When a bin is empty the worker signals his need for further supplies. Everything else happens automatically, initiated by the relevant information system.

The container is brought to the outer area of the MSS by the driverless transport system. There the empty compartment of the assembly container is loaded with a full replacement bin which is taken from the storage alley by the rack operating device and placed in the assigned container compartment.

After loading by the MSS, the driverless transport system delivers the smaller parts via the assembly containers, and the larger parts on palettes or in grid boxes to the workplace.

To ensure that every part is carried to the correct assembly place, and from there to the correct delivery point the individual assembly places must be precisely described in computer-technical terms - much more precisely than before, when supplies to the assembly locations were still handled by manual transport processes.

For this purpose each assembly location is broken down analytically. From the **bills of materials** it is determined which parts/aggregates are assembled at the workplace.

From the layout of the workplaces the individual positions for the placing of containers or grid boxes can be determined (see Fig. D.I.11). These locations are provided with addresses in accordance with the above system. Thus, each container/grid box receives a supply location address, which is linked in data-technical terms with the container number. The loading of each container with the various parts needed is also described in data-technical terms.

3. Integration of the Automated Logistic System and Production Planning and Control (PPC)

As has been shown, a multiplicity of feed lines or vehicles on a fixed transport line need to be set in motion for the automated transport system (constant feeder and driverless transport system). Furthermore, rack operating devices and perhaps commissioner robots need to be put into motion. For this purpose 95 controls are employed.

Control computer systems are required for initiating and monitoring the material flow. For this purpose 6 control computers are employed, which supervise the 95 controls. In the framework of the overall system these control computers take over operative functions. They are not independent executive systems, but are embedded in a data and program environment along with PPC procedures on the HOST computer (see Fig. D.I.12).

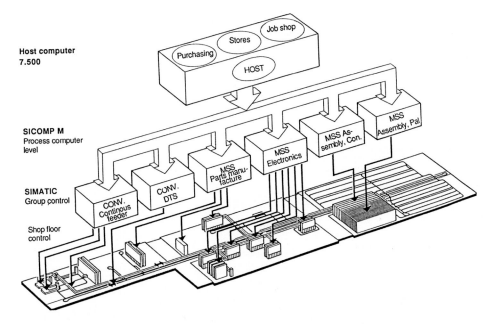

Fig. D.I.12: Control computer system for transport and material supply

The following procedures in particular are to be integrated on the **HOST computer**:

-   **purchasing procedures,**

    because incoming orders and their destinations are stored there,

-   **stockkeeping procedures,**

    because the entire stock management and planning reside there,

-   **job-shop control procedures,**

    because production orders which give rise to transport processes are stored there.

4. Integration Axis: Databases and Programs

The factory needed a new database to incorporate the integration areas discussed. This function was assigned to the **factory primary database** for the **static data** and to the **container database** for the dynamic data which monitor the container movements (see Fig. D.I.13).

The essential additional data of the new primary data system are outlined by way of example. First, one recognizes the "classical" data of primary data systems, such as product descriptions (parts master file), bills of materials, work schedules and workplaces.
Each of these classical data areas was extended with essential additional data. For example, the previous **parts master files** were extended to include additional details relating to

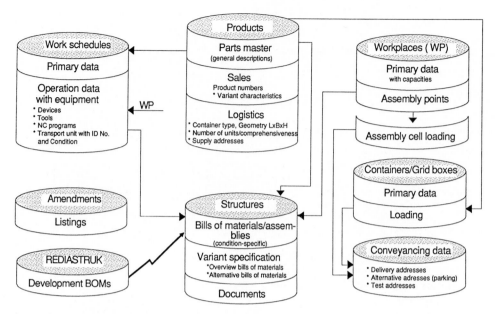

Fig. D.I.13: Database of factory primary data

container type, the suitable number and dimensions for processing, and the supply address to which the parts should be delivered.

The **work schedules** specify the equipment needed for production - such as devices, tools, NC programs, transport units - covering a much greater range than previously.

The new data content is particularly obvious for the **workplaces**. In an automated factory in which material is delivered to centimetre accuracy completely automatically, the workplaces need to be described in meticulous data-technical detail.

This is particularly clear for the **assembly workplaces**, where each assembly island must be described exactly with the location of containers, grid boxes, and palettes as well as the individual compartments in each location. In this way the new primary data system represents a decisive mean towards the integration of the complete CIM system.

5. Complete CIM System

The complete CIM system described here requires the linking of **heterogeneous computer systems**. The systems in the **PPC** environment and the **CAE** area are mounted on **HOST computers**.

The **production control system** (not presented in this article) and the container information system are mounted on **departmental computers**. The executing system for transport and material supply (MSS) as well as the control computer for the automated

219

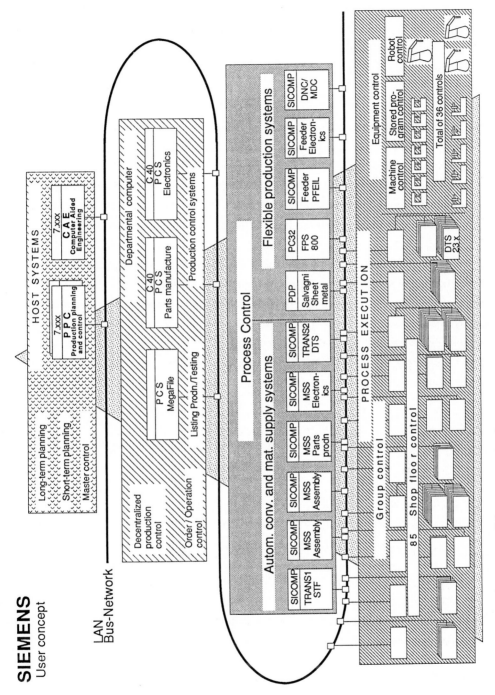

Fig. D.I.14: Total overview of the integrated information systems

electronics assembly line and the DNC systems for mechanical manufacturing are **process computers**. At the flexible processing system level PCs are generally installed. The physical execution of movement and processing procedures are handled by controls. In the new factory the various systems are linked together using several local networks (LANs) (see Fig. D.I.14).

Distributed Processing and Distributed Databases:
The complete CIM system presumes the completely integrated use of the various procedures on the various computers. The computers, and hence the programs, are distributed spatially: the mainframes are located in the computer center, the departmental computers are installed decentrally in the factory halls, as are the process computers for the automated transport, inventory and processing procedures, the PCs are, of course, distributed to the workplaces.

In this context reference must be made to another important factor. The complete integration of the systems necessitates the highest degree of integration, that is **program-to-program link**. Since a procedure may begin in any one of these linked processes, and can on occasion extend over several other processes before it is completed the linking of the various processes must occur at the procedural level.

This high integration requirement - the **online** linkage of all relevant processes - demands the implementation of distributed data processing and the use of distributed databases.

f. HP OpenCAM - Realization of a CIM-Strategy for Production Control with ARIS (Architecture of Integrated Information Systems)

(Prof. Dr. August-Wilhelm Scheer, Dipl.-Ing. Wolfgang Hoffmann, Dipl.-Wirtsch.-Ing. Ralf Wein, Institut fuer Wirtschaftsinformatik (IWi), University of Saarland)

1. The HP OpenCAM Approach to Problem Solving

The constantly growing demands regarding flexibility, productivity and quality move today's companies toward new challenges. These mainly consist of responsibilities to build and maintain the companies competitiveness. One way to meet that challenges is the CIM-strategy (Computer Integrated Manufacturing) which comprises the integrated application of information technologies for managerial and technical concerns within a company. The realization of integrated information systems is mostly be accomplished by different partners. These may include computer manufacturers, consultants, experts from various departments of the company or EDP-specialists (Electronic Data Processing). The system developed by these partners has to accomplish universal requirements concerning organizational and EDP-specific aspects. Therefore each of the partners must use the same set of rules. As the demand for integration increases the significance of this becomes more and more important. An example is the unified support of business processes.

Within production control the integration of assembly control, production data collection, quality management and technology is the most important issue. If EDP-systems are developed as isolated solutions, they may provide a high standard support within their specific functions. However, regarding the production process, they lose in importance. This can be observed on the supply side of today's production control EDP-systems. Most of them provide a high standard of functional usability, but the degree of integration between different systems is at its lowest. Facing these problems Hewlett Packard has developed a concept of open systems: HP OpenCAM. The idea is based on standardized "best-in-class" solutions by innovative HP-partners in the following areas:

☐ production control: intelligent shop floor control system FI-2 developed by IDS Prof. Scheer GmbH,

☐ production data capturing: Factory Integration Tool (FIT) developed by A & B Systems,

☐ quality assurance: SyQua developed by EAS,

☐ NC-programming: UNC 8500i developed by UNC,

☐ process analysis and control: APROL developed by PLT.

In order to guarantee the functioning and co-operating of these five CIM-components an appropriate description method for the systems had to be found. Besides the sole description, it needed to analyze the integrative patterns of the systems and had to classify them according to enterprise-wide processes. The set of rules according to which the universal description of information systems is possible can be viewed as an architecture. This architecture allows different individuals involved in the process of developing an information system to refer to a common basis.

2. Classification of HP OpenCAM with ARIS

Previously EDP-systems were developed and introduced too early, the integration of the substantial correlations often lacked sufficient analysis. In order to guarantee a maximum of integration with all needed applications, the Architecture of Integrated Information Systems (ARIS) is applied for HP OpenCAM.

The components of an information system, which is described from the business application standpoint, including their relationships to each other, are conditions, events and processes; the factors of production are: material, human labor (employees), equipment and the organizational units:

- ❑ the fundamental elements are process and event. A process is a unit of some duration which is started by an event and completed by an event. Thus, start and result events define the beginning and end of the process. With these elements the sequence of processes and their dependencies on events can be described.
- ❑ Materials which are needed within a process as well as materials which belong to the output.
- ❑ All information, which is needed for the process and which will be amended by it, is grouped together to conditions. E. g., conditions from the environment must be included, which provide parameters for the processing rules. While processing these data can be altered, for example, inventory can be reduced by assigning components to the customer order.
- ❑ Human labor can be grouped together in organizational units and also technical equipment which can be production equipment as well as EDP equipment.
- ❑ In general, the links between all the elements is defined by arrows. Every element may be used more than once per process. Not all arrows are equal. Depending on the type of the linked symbols, they have different meanings. The arrows connecting events and processes define the control flow. Those between the processes and the conditions imply an information flow. Material and processes are linked by the material flow arrows. The arrows from the process to the human resources and those to the equipment define the support of the remaining factors of production.

These are the elements of business processes. In the first step complexity is reduced by considering only those facts which are relevant from the information transformation view. But the process remains still very complex.

The concept of ARIS shown in Fig. D.I.15 consists of two different parts, a method to break down complex processes into their components, and a method to define different phases of description. The first is based on the idea that business processes are by far too complex to be handled as a whole. Rather, they have to be subdivided into their components, if planning and realization of information systems is to be successful. These components have to represent all possible views of an information system as well as all relationships between them. As that views ARIS defines data, function and organization view connected by the control view.

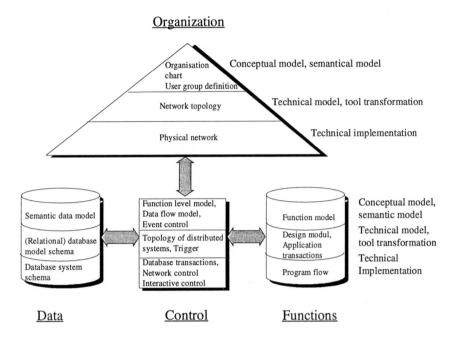

Fig D.I.15: Architecture of Integrated Information Systems (see *Scheer, ARIS 1992*)

Conditions, events and relevant environment conditions are represented in the data view. The different units of data and their relationships are handled as single information objects in the data view. The description of complex functions related to the processes within a company is accomplished in the function view. Relevant functions are depicted in a static and clear structure, from which subordinate and superior functions emerge. The organization view focuses on the users and organizational units. Their specific and disciplinary hierarchy and as well as their structure are described. Due to the division into three different views the relationships between that views have to be represented by a fourth view. The control view describes basic relationships between the data, function and organization view. By using the architecture it is possible to describe each view in an isolated way. Their relationships are presented in the control view. The different phases of description are the requirements definition, the design specification and the implementation description. The requirements definition models the individual viewpoints of the information system independent of following implementation considerations. Here, descriptive languages which allow a precise definition are used. The results are semantic data, function organization and process models. In the next step the semantic models are adapted to the demands of user interfaces and hence transferred into design specifications. The design specifications may include network topologies, database systems, programming systems or trigger mechanisms. The third step is the technical implementation description by adjusting the design specifications to concrete information technologies. Physical networks, physical data structures, hardware components, executable program codes and the allocation of computer resources result (see *Scheer, ARIS 1992*).

The development of the three description levels does not run in the same way parallel (see *Scheer, Communication Technology 1991*). While the implementation description changes in short periods due to the fast development of new information technologies the requirements definition with its contents and conceptual requirements keeps its validity a long time. The design specification is only amended if a basic and technical redesign of the system is necessary due to the numerous cycle of information technology. The significance of the requirements definition becomes obvious. For that HP OpenCAM concentrates on the requirements definition level.

Contrary to the static nature of the semantic data, function and organization view, the semantic control view describes a more dynamic view within an information model. The semantically process model of the control view represents the time and logic dependent of functions.

This are the ingoing and outgoing data of a function and trigger mechanism which starts functions. The trigger mechanism can also be referred to as an event and is characterized by modifying the instance of an information object within the data view. Events initiates functions and can also be results of a function. They refer to on attributes which belong to the information objects of the data model. Consequently, an event describes the taking place of values of attributes which initiate functions. Due to that there is a relationship between the events of the process model and the attributes of information objects. An event can refer to one or more attributes of an information object, and an attribute of an information object can relate to one or more events. As far as a complete data model exists it is possible to identify information objects and to analyze possible values of potential events. If a data model does not exist, significant practical events are to be identified. The complex correlations described above can be represented by a graphic. To show the mentioned relations event driven process chains (EPC) can be used. The EPC describes which events initiates which functions and which events are generated by which functions. Since an event generated by a function, also triggers a new function a coherent chain results. Between events and functions links can be defined, which describe the relationships. Mostly the resulting chain has a reticular shape. Fig. D.I.16 shows an example of an EPC for HP OpenCAM.

Immediately after checking the availability of all resources necessary for an operation order by the Leitstand they will be reported back by subsystems. This is the starting event of the EPC shown in Fig. D.I.16. The Leitstand verifies the completeness of resources,

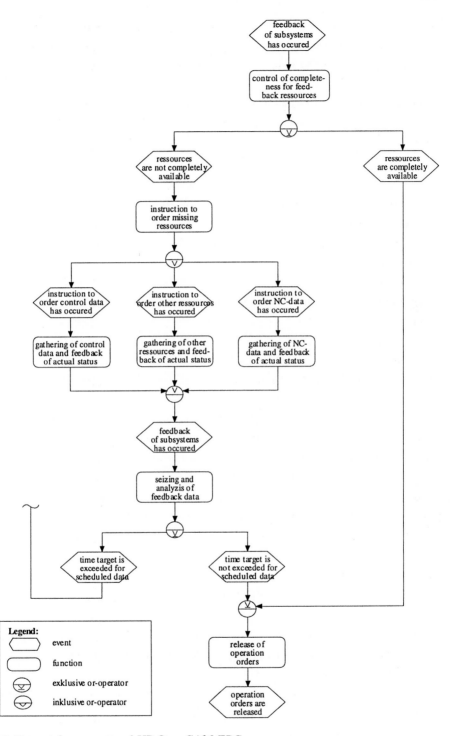

Fig. D.I.16: Extract from a general HP OpenCAM-EPC

reported back by subsystems. If all resources are available, the operation order is released. Otherwise, if resources are not available, the Leitstand instructs the subsystems to provide resources necessary. The subsystems will report the current work status of resource provision to the Leitstand. The Leitstand collects and analyzes the data provided. If the planned time for the scheduled data of the Leitstand is not exceeded the operation order is released. Otherwise a rescheduling of the earliest date of availability of the resources is necessary.

3. Transparency of the System-Architecture of HP OpenCAM with a Hypermedia Documentation Tool

Each of the rough process models are based on more detailed process models of particular systems. The obviously high documentation volume of ARIS leads to the need of using an appropriate EDP-system. For this reason the "Institut fuer Wirtschaftsinformatik" at the University of the Saarland, Germany, has developed a system called *ARIS-Navigator*. The *ARIS-Navigator* is a hypermedia navigation tool with an object-oriented user interface. It supports the users to navigate through all description phases and views of ARIS with a universal structure for information retrieval. The *ARIS-Navigator* allows the selective presentation of information and view-specific preparation of complex information models. Within each information model all data stored in the underlying ARIS-Repository can be retrieved (see *Nüttgens et al., Information Controlling 1992*).

For HP OpenCAM all four ARIS-views were modeled in three different layers of detail and were then implemented into the *ARIS-Navigator*. The open structures of HP OpenCAM in the *ARIS-Navigator* allow potential users an easy integration of the concept. An example of the user interface of HP OpenCAM in the *ARIS-Navigator* is shown in Fig. D.I.17.

4. Standards used for Realizing HP OpenCAM

HP OpenCAM is built as an open, modular and flexible integrated CAM-system that allows a synchronized implementation in order to obtain an optimum reconciliation with the individual situation of users. The first step to reach that goal was to define the given technologies and applications for the integration platform in order to merge the different CAM-systems into one unified integrated CAM-system.

Standards of the user interfaces were:

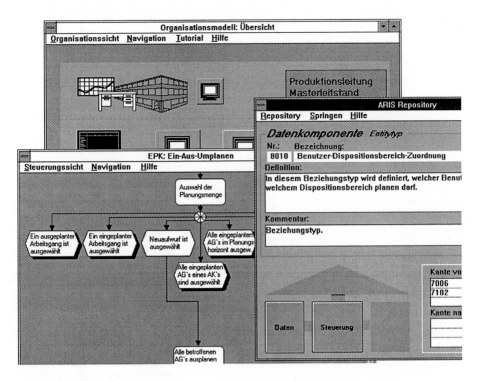

Fig. D.I.17: Example of the user interface of HP OpenCAM in the *ARIS-Navigator*

- ◻ OSF/MOTIF (look and feel),
- ◻ X11 (networking),
- ◻ XTOOLKIT (graphics),
- ◻ XLIB (graphics) and
- ◻ PHIGS/PEX (graphics).

For data storage and administration SQL-databases (ANSI level 1,2) are used. The communications networking services used are IEEE 802.3, Ethernet, TCP/IP, ARPA/Berkley Services, NFS and SNA. Based on these standards and application systems an interface specification with a neutral file-format has been developed. This allows the integration of new systems on the basis of the standards mentioned above into HP OpenCAM in the future.

5. Principles of the Interfaces

Concerning the interfaces within HP OpenCAM the underlying principle is that of "dept at the others domicile". Interface programs are generally activated within shellscripts. Fig. D.I.18 illustrates the principle flow of data transfer.

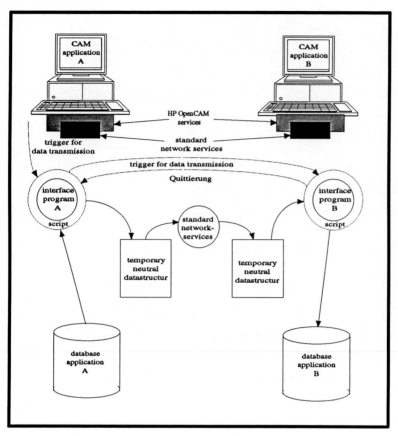

Fig. D.I.18: Flow of data transfer between applications

If application A wants to send an information to application B, a script is activated by A. Within the script an interface program is started to read the information from the database and to transform it into a temporary and neutral data structure e.g. a file. The data that are to be extracted for the target system can be adjusted by the interface program in order to deliver only the necessary information. The data is transferred to application B by commands within the script. Once at application B the script of application A activates the interface program of B. The interface program of B then writes the information into B's database, discharges the successful processing and deletes the temporary file afterwards.

Errors occurring during the information flow are reported to a file, that can be visualized if necessary.

This enables the systems of HP OpenCAM to communicate in one "common" language, which corresponds to the integration in the sense of CIM (Computer Integrated Manufacturing).

6. Conclusion

The idea of CIM will only become reality, if open systems, as a part of a superior architecture can enable available and future applications to be integrated into that platform. The concept of HP OpenCAM by Hewlett Packard points out future guidelines which combine the power of open systems with the high competence of applications, and with their flexibility.

g. Computer-supported Reengineering of Business Processes: Requirements for Successful Lean Management

(Prof. Dr. A.-W. Scheer, Institut fuer Wirtschaftsinformatik (IWi), University of Saarland)

1. Introduction

The automation of existing business processes alone seldom leads to the highest gains in productivity. Experience in the areas of Computer Integrated Manufacturing and Computer Integrated Office have shown that a narrow focus on automation primarily leads to frustration. The key to success often lies in substantial modification and improvement of the business processes themselves which is supported by a fine-tuned information system. In this situation, the term lean manufacturing, is increasingly used to characterize a path for improvement, which emphasizes the need to focus on the business activities. The concept of lean manufacturing was coined in a large MIT (Massachusetts Institute of Technology) study which compared Japanese, European and American automobile companies (see *Womack, Jones, Roos, Automobile 1992*).

Lean manufacturing requires new organizational structures. The goal is to reduce organizational hierarchies, to replace the tayloristic division of labor by re-integrating functions, and to significantly increase the amount of team-work.

Strategies for lean management can profit from recent developments of computer science. Although traditional approaches for the implementation of EDP systems have often neglected the organizational structures and real business processes into which the EDP systems should have been embedded, this deficiency has recently become obvious and new approaches have been developed to systematically derive an implementation specification starting out from a description of the business processes. An example of such an approach is the Architecture of Integrated Information Systems (ARIS) (see *Scheer, ARIS 1992*) which shall be presented briefly and then is used to discuss alternative approaches to structuring an organization. The function-oriented approach to business process organization is criticized and a case is made for improving business processes through integration, simplification and decentralization.

2. Architecture for Process-Oriented Information Systems

An architecture has to reflect the main components of an information system. The most important business perspective is on process chains, which consist of a sequence of business processes. Organizational structures, upon which lean manufacturing focuses, and information systems have to efficiently support these process chains. However, a systematic analysis of process chains is difficult because of their complex interrelationships and dependencies. The complexity can be reduced by introducing views upon the business processes which group certain types of information. In the function view, a set of business processes is decomposed into a hierarchical structure of functions. In the data view, the classes of information which trigger processes, which are required for the execution of processes or which are created by processes are defined including their relationships. In the organization view, the relationships between organization units that are involved in business processes and responsible for information and functions are described. These different views are integrated in the control view which links functions, data and organization to a process and which allows to describe a sequence of processes, the process chains. To separate information concerning the business from information concerning the technical support, a separation into three different layers, the conceptual layer, the technical layer and the implementation layer has been incorporated into the architecture. The conceptual layer captures the description of business process independently from any technical considerations. The two lower layers are used to derive the technical implementation and are not of further concern here. The Architecture of Integrated Information Systems (ARIS) has been derived in this way. A figure of ARIS is shown in the article of HP OpenCAM. ARIS can now be utilized to analyze and subsequently optimize the business processes of a company.

3. Alternative Approaches for Business Process Organization

The effort required to solve a problem often is proportional to its complexity. The complexity of an organization depends upon the number of functions and the number of objects which have to be managed or to be produced. Both are parts of process chains: order acceptance, materials requirements planning, capacity management, production control and shipping are part of the production logistics chain. Product development, design, generation of work sheets, NC-programming, testing, and quality assurance are part of the product development chain. The interdependencies between functions can be classified as data and decision relationships. A data relationship exists if a function uses data from another function. A decision relationship exists, if a decision taken in one function is also influenced by decisions taken in other functions. A second characteristic for the complexity is the large amount of objects to be managed, for example the products, assemblies, individual parts etc. Between objects interdependencies also exist. If attributes of an object are changed, for example the modification of the completion date for a product, this might also lead to changes in other objects, for example the modification of the completion date in a subordinate part.

The high number of interdependent functions and of interdependent objects therefore are the source of complexity in an enterprise. This is shown in Fig. D.I.19 by the size of the rectangle. The complexity is so high that no solutions for management and planning are known, which can simultaneously take both dimensions into account. Instead, the complexity is reduced by dividing the problem into smaller problems.

The main approach has been a functional division of process chains. Large enterprises are divided into functional units which are responsible for purchasing, production, sales etc. Functional units again are functionally structured into departments which then might again be functionally structured into groups. The result of this division is shown in Fig. D.I.20. Horizontal blocks indicate a functional area. The functional division leads to barriers between the departments and to sub optimal solutions. A good approach to reduce the negative effects of this division is to ensure, that the functions are efficiently linked. This requires the possibility to exchange data between different functions and the certainty that data is correctly interpreted in all the functions. Many organizational measures, such as

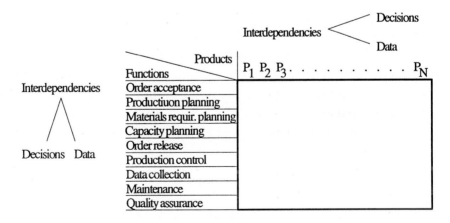

Fig. D.I.19:Interdependencies in a manufacturing enterprise

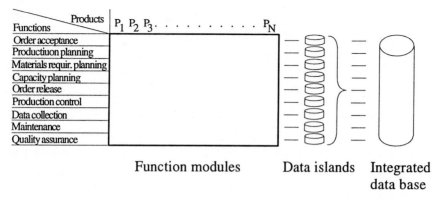

Fig. D.I.20: From Data islands to the integrated database

consulting groups, quality circles, notification procedures etc., have been developed to support this information integration. With computer-support, it can be accomplished by using an integrated database, which is shown by the cylinders of Fig. D.I.20. This information integration, however, may again substantially increase the complexity, as many interdependencies between functions have to be monitored.

4. Integration, Decentralization and Simplification

The disadvantages of strictly functionally structured organizations have increasingly been pointed out in the last years. Multinational companies often have divisional structures which corresponds to a rather vertical segmentation in Fig. D.I.19. Product oriented segmentation is carried out by identifying functions which have to be carried out for a family of products. The assumption is that the dependencies and relationships between the associated processes of different segments is much smaller than the dependencies between functions which belong to the same process in one segment. In the area of production, this has lead to the development of concepts for flexible manufacturing systems and production islands. Within a production island a family of similar products is produced. All manufacturing operations required are carried out in the production island. The object-oriented organization of production islands has the advantage of significantly increasing throughput time in comparison to the traditional job-shop environments. However, capital costs are higher as machine capacity tends to be less efficiently used.

The production island principle currently is discussed or already applied in many companies. For every decentralized production island, a higher degree of autonomy, skill and responsibility for a range of different functions has to be provided. However, this increase in power is limited by a small number of objects, to which the functions may be

Capacity planning					
Order release					
	Coordination				
Production control Data collection Maintenance Quality assurance	Production island 1	Production island 2	Production island 3		

Fig. D.I.21: Reduction of complexity with coordinated production islands

applied. Therefore the increased complexity concerning the functions is offset by a decrease in the number of objects to be managed. This is shown in Fig. D.I.21 where the functions integrated within the productions island comprise production control, quality assurance, transport management, NC-programming and maintenance. Between different production islands, interdependencies may exist, because the parts produced in a production island are sometimes assemblies, which may be combined to form more complex products. Interdependencies can also result from sharing resources between different production islands. These relationships have to be managed by introducing a co-ordination level between the different production islands. Such a concept has already been implemented in production control software (see *IDS, Leitstand 1990*).

The approach which has been described using the production island as an example can be applied to all parts of the enterprise. The throughput times of the production logistics chain can be decreased by integrating the functions for sales, requirements planning and capacity planning. The increased complexity of the function again is offset by product-oriented segmentation. The same principle can also be used to improve the product development chain. By integrating the functions of product design, tool design, calculation, NC-programming and testing the concept of simultaneous engineering can be realized. Also in this case the increased complexity of the functions is followed by a reduced set of objects to which the functions apply. This approach is shown in Fig. D.I.22 as co-ordinated decentralization. Two aspects are of primary importance: the integration of functions within each of the decentralized units and the co-ordination between the different units, which ensures, that problems which affect several units can be efficiently solved. Synergy's between different units therefore are not lost.

The process chain as basis for building an organization structure increasingly is used to improve inter-company co-operation. An example are the close links between customer and supplier in just-in-time environments.

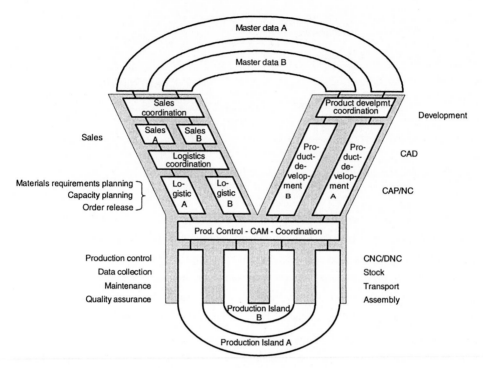

Fig. D.I.22: Coordinated decentralization in the Y-CIM-Model

5. Conclusion

The concept of co-ordinated decentralization is an important application of lean manufacturing. It increases the autonomy of the employees by re-integrating functions, it reduces the effort for planning and co-ordination and reduces the number of hierarchies in the enterprise.

Although experience with lean production is available from Japan and from new production sites in the United States, this concept often is advocated in very general terms. For successful application and for the implementation of the concept, however, detailed methods and approaches have to be devised. The general concept of lean production has to be solidified with concrete organizational concepts and implementation strategies. New concepts from information technology are well suited to give support in this endeavor. Lean production therefore does not oppose the increased application of computer support.

The concepts developed for modeling processes and data are very well suited for this type of application. As the introduction of lean production leads to increased functional integration, processes are accelerated and have to be redefined. The definitions of the corresponding functions have to be modified and their interrelationships with predecessor and successor functions have to be updated. The same applies to the allocation to organizational units, for example for newly built teams. These problems are very suitable for modeling. Modeling also has the advantage that it can be used for much more than for documentation. It is more powerful than the traditional organization handbooks, because models can be efficiently visualized, easily distributed and much easier maintained. In addition, computer-based modeling provides the possibility of quick analysis (see *Jost, Werkzeugunterstützung 1993, Nüttgens, Scheer, ARIS-Navigator 1993*) and even simulation of business processes, which is a prerequisite for their optimization. Slightly further in the future is the most formidable application of information modeling: models of an enterprise may be used not only to evaluate the suitability of software for the enterprise but also for the configuration of existing software. Then business process optimization and the design of an information system will truly be what they should: two perspectives on the same issue.

II. CIM Solutions in the USA[1]

The enterprises or factories described
- IBM's factory in Lexington (Kentucky),
- HP factories in Lake Stevens (Washington) and Cupertino (California),
- LTV Aircraft Products Group in Dallas (Texas),
- Westinghouse Electrical Corporation in College Station (Texas),
- Digital Equipment Corporation in Springfield (Massachusetts),

are certainly not representative of all of American industry, but they demonstrate the feeling of change that is being aroused in large sections of industry in America by the CIM concept. The works considered exhibit diverse accents in production: IBM Lexington manufactures electronic peripherals with a strong assembly orientation; HP manufactures medical measuring instruments and computers; LTV manufactures parts for aircraft construction using metal cutting (milling) procedures; Westinghouse assembles electronic circuit boards at College Station; and DEC in Springfield manufactures a wide range of fixed disk devices within an assembly-oriented production environment. The focal CIM issues within the works are correspondingly diverse.

a. IBM Works, Lexington (Kentucky)

The IBM works in Lexington (Kentucky) are already regarded as the classic example of CIM. Although, in the meantime there are more impressive examples within the world-wide group of factories (also in the IBM works in Germany, for example) this factory remains interesting

---

1   These descriptions are the result of several study trips by the author in the years 1986-1988 to American research and development laboratories as well as visits to several CIM factories. A more extensive treatment is given in *Scheer, CIM in den USA 1988*

on account of its radical changeover to CIM. The history of the works is therefore a typical example of how CIM can change the enterprise strategy in the industry.

The factory, which was founded in 1956, produced electric typewriters up until the beginning of the 1980s. By the end of the 1970s the strong competitive pressure from Europe and Japan as regards price and quality was already becoming a problem: whereas the number of competitors in 1979 was only nine, this had grown by 1986 to 40. This gave rise to the alternatives of either abandoning the typewriter market or tackling the competition with new products and production techniques. The recommendations of a task force were presented in 1981 and proposed a more strongly electronic product and the simultaneous introduction of highly automated production. The restructuring of the factory, which could only be completed after six years, cost around 350 million dollars. The number of employees fell in this period from 6,500 to 5,000. The manufacture of the traditional products was continued in parallel with the development of the new product lines.

The goal was specified as one of producing the new products at a third of the previous price, while at the same time increasing the functional scope and improving the quality of the products. To do this the number of mechanical parts was drastically reduced, and the multiplicity of models limited to seven essential product types. A strict product group technology allowed a high proportion of parts with multiple uses. The current production program consists essentially of electronic typewriters, printer typewriters and printer workstations.

Although when the factory was being developed a self-contained CIM philosophy, such as is found in modern textbooks, was not yet available, essential fundamentals were already implemented. These are:

- early involvement of production in design and development,
- transition from job-shop production to process production,
- drastic reduction in inventories.

Inventories, graded according to part type, are held with a coverage of between 20 days and less than one day. A comprehensive "just in time" provision is not pursued, but rather a "realtime delivery" policy in which the preferential rates on large order quantities for parts of low value can be exploited.

Low inventory coverage necessitates greater quality reliability from suppliers. This is achieved through close cooperation between the factory and its suppliers. Around 80% of goods received are no longer subject to testing at receipt of goods, since this has already been carried out in the final quality checks of the supplier. The number of suppliers was initially reduced from 1,000 to 700; a further drastic reduction to around 60 suppliers is planned. This also has the effect of reducing the administrative costs which previously arose in dealing with a large number of suppliers.

To support strict automation, highly automated manufacturing facilities are employed for the pre-fabrication of parts. 154 of IBM's own robots are used in assembly.

The hardware situation is characterized by two system environments: Systems from the 370 architecture are used for planning functions, whereas around 170 S/1 process computers with the operating system EDX are installed for production control (see Fig. D.II.01).

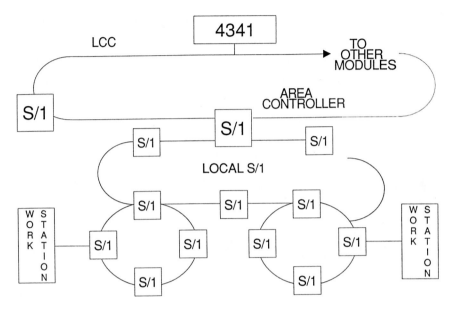

Fig. D.II.01: Computer structure in IBM's Lexington works (USA)
Source: *IBM*

The automation effects can be indicated by comparison of the cost structures. Whereas in 1982 the relative proportions of overheads, pay-roll costs and material costs were 42%, 10% and 48%, by 1986 these proportions had changed to 18%, 5% and 77%. The streamlining of processes resulting from strict process chain analysis along with the reduction in inventories has especially reduced the indirect and pay-roll costs.

Analysis in terms of process chains is supported by a new organizational structure based on the object principle. The works manager is assisted by product managers who are each responsible for design, development and production within their respective product programs. Communication between the product areas is effected via:

-       a common technical database for parts and work schedules,

-       the employment of identical computer tools (CAD systems), and

-       the use of the same production units.

In total, 5.500 finished products leave the works daily.

238

The S/1 series of computers are embedded in a CIM architecture which is defined for a departmental, production cell, and equipment level (see Fig. D.II.02). This accommodates the functional architecture of departmental tasks, cell functions and equipment control. The control software was largely developed in-house. The concept that was developed in the process currently serves as the basis for the design of new standard software in Boca Raton. In order to accommodate the diverse requirements of the various assembly systems and production units it was decided not to develop a rigid control software, but rather a kind of toolbox to support the development of dedicated systems. Of special importance in the toolbox PACS are processors for communicating with diverse machine controls and for communicating with other system entities as well as for the administration of production data. The system PACS serves as the model for the toolbox developed in Boca Raton with so-called "enablers" for developing CIM applications software.

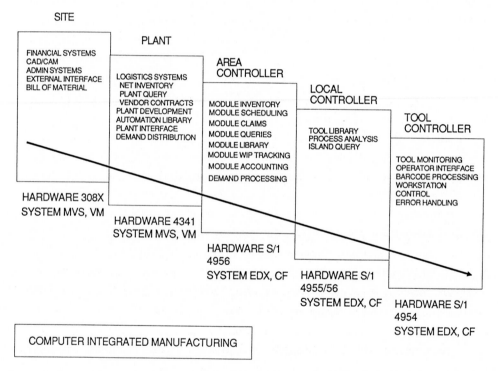

Fig. D.II.02: Functional breakdown at the hardware level in the IBM Works, Lexington (USA)

In addition to production control software, simulation models exist to optimize the production layout within the factories, and to recognize production bottlenecks.

b. HP Works in Lake Stevens (Washington) and Cupertino (California)

The high proportion of material costs in total manufacturing costs within the electronics industry (the proportion varies between 45% and 70%) explains the great significance attached to an order logistic aimed at reducing inventories and throughput times. In both the works for medical measuring instruments in Lake Stevens and at the works in Cupertino, where computers of type HP 3000, HP 1000, and of the new Risc architecture are manufactured, the simplest possible planning procedures are aimed at optimizing the material flow. Although the two works produce different products, close similarities exist. In both factories a low processing intensity allows a reduction in the complexity of production control. In one production area electronic circuit boards are produced which are then mounted on a chassis in an assembly-oriented area to create the end product. The low processing intensity does mean, however, that the more complex production control problems are relocated to the preceding suppliers. In the works themselves, however, it is possible to achieve considerable success by modest means. This fits with the motto that if the organizational prerequisites for a CIM-competent factory are achieved the subsequent improvements in production technology come in its wake.

Given these preconditions it is understandable that standard software systems for production planning and control designed by HP (MM/3000 and PM/3000) for the high processing intensity that is normal in manufacturing industry are not suitable. In Lake Stevens the available MM/3000 system was therefore extended using in-house developed systems or partially by-passed by immobilizing program code or exploiting program exits intended merely for special cases.

The starting point for the development of a Kanban system in Lake Stevens was the fact that the lots based on weekly requirements of the preceding "push" philosophy had led to long throughput times and high inventories. Furthermore, the constant changes in the master planning system of production planning generated considerable disturbance.

The Kanban system was developed in three stages.

In the first stage the production orders generated by the MM/3000 system were merely used for inventory commissioning, so as to prepare the components needed for the production of circuit boards and the assembly of the end products. Within the production line, however, the material flow was controlled by Kanban cards. This allowed a drastic reduction of inventories held within the production line. Lot sizes within production were also considerably reduced. Although no changes were made with regard to the PPC system (all information was printed out as before) this was avoided in practical terms, since the working papers were not used for production control.

In the second stage the output of working papers from the PPC system is completely eliminated. Commissioning from stores is also effected via Kanban cards in accordance with the "pull" principle. The program exit "unplanned withdrawals" serves as the interface with the PPC system.

The "requirement breakdown" function of the PPC system is used only to establish the need for purchased parts, but ceases to have any influence on the generation of production orders. The erosion of the classical PPC system by the Kanban system is extended by an in-house developed master planning system which supports medium term production planning using graphic support on a prior Lotus-1-2-3 system. This system is the starting point for the numbers of bought-in components required, determined with the help of the requirement breakdown. The statistical monitoring of incoming customer orders is checked constantly to determine whether the forecast values used are still accurate or need to be corrected.

The changes in the weighting of the functions, as compared with traditional PPC systems, which support master planning and short term production control as opposed to the medium term requirement breakdown, indicate a more suitable PPC philosophy for forward-looking CIM structures.

The successes of the system are compelling: the throughput times for end products were reduced from 45 days to between four and five days; in addition, the production times for circuit boards fell from between 16 and 20 days to one-and-a-half days. In Cupertino the throughput times were reduced from a previous two weeks to a current 8 to 36 hours.

Whereas in Lake Stevens the Kanban cards are produced with the help of a computer system and destroyed once the restocking process has been carried out, so that errors arising from technical changes are avoided, the system in Cupertino is more robust. Here, the material flow is partially indicated using colored markers on planning boards. The number of green magnetic counters indicates how many Kanban units produced are currently available, the number of red markers indicates how many Kanban containers have already been transferred to the next production level, so that by simple addition the total number of units produced is constantly visible. Out-of-stock-processes are shown on the planning boards by the controllers, so that these are constantly visible to everyone and thereby minimized by the resulting social pressure.

The third stage of development of the Kanban system was dispensed with at Lake Stevens. These entailed plans to involve suppliers directly in the supply logistics by using a just-in-time approach, so that stocks of goods received could be largely eliminated. The additional economies achievable in the third stage, however, seem, in comparison with the improvements already achieved, too small to justify taking this step at present. In contrast, Cupertino has already introduced a just-in-time philosophy for its suppliers in the assembly of the computers. The computer chassis are delivered directly from the lorries to assembly, in order to save the space needed for these bulky parts in the storage areas.

In the works great value is placed on consummate quality assurance. This already starts during development and is continued through production: "For every dollar that is invested during research and development to develop the quality of a product, we save ten dollars in production."

In Cupertino there is no special testing department; quality checks are carried out within the production line after every processing step, so that only error-free components are passed on to the next processing step.

The high level of organizational decentralization, which is associated with relatively autonomous planning, should not obscure the fact that the enterprises are controlled by a tight network of norms and monitoring. An attempt to solve the difficulties in evaluating CIM activities using traditional economic efficiency analyses, for instance, is made by using a host of quantitative indicators.

Each organizational level has the following management resources:
- development of a strategy,
- development of target figures,
- development of a catalog of measures for achieving the strategy,
- evaluation of the measures.

These are made operational by the use of index numbers.

For example, in Cupertino the goal of making greater use of product group technology was made operational by adopting the norm of increasing turnover in each assembly group by 50% in three years. In creating the software the aim is to reduce maintenance costs from their present 40 dollars per line of source code per year to 25 dollars.

The essential starting point for HP's CIM concept is the organizational levels model presented in Fig. D.II.03.

The organizational levels company, plant, plant area and production cell are assigned typical functions (application areas). The functional assignment is based on the processing range that the function covers. If it is applied to the entire company, it is assigned to the highest level, conversely, if it relates only to a single production cell it is located at the lowest level. The typical time intervals assigned, with the tendency towards increased currency in the direction of the production process, can only be followed up to a point, since, depending on the assignment of functions, even at the company level temporally

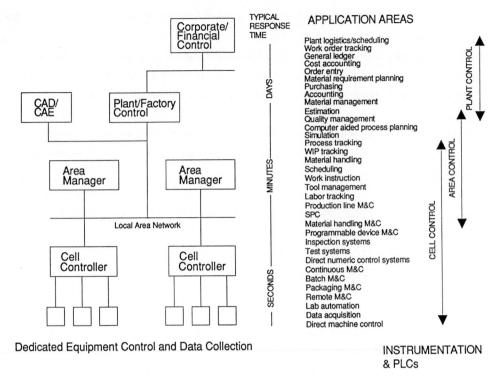

Fig. D.II.03: Production environment
         Source: *HP*

critical processes can be due. Nevertheless, the organizational functional architecture is a helpful starting point for the further steps in CIM planning.

c. LTV Aircraft Products Group, Dallas (Texas)

LTV had already installed a highly automated production cell in its works in Dallas in 1984. The great efficiency this achieved provided the impetus for the currently almost completed integrated flexible production system IMS (Integrated Machining System, see Fig. C.VI.03). The physical space occupied by IMS is 100,000 sq. ft. Extremely large aluminium and titanium aircraft parts are produced in milling manufacture, whereby considerably more than 1,000 different parts are produced using the most modern high speed machines.

These include five-axle milling machines of heavy construction with automated tool supply, material parts supply via driverless transport systems and storage systems. The entire system is monitored from a central computer room with three VAX Cluster systems and numerous peripheral control systems. The system was created by Ingersoll (USA) as the main contractor, whereby equipment from Cincinnati Milacron and controls from Allen

Bradley with Zeiss checking devices, among others, are linked together. Alongside the integration of CAM components the foreground is occupied by the system's material supply, which is effected via a computer controlled logistic model. The importance of the logistic is emphasized by the description of the Job Service Center as the "backbone of the system".

IMS is part of a comprehensive CIM strategy with around 15 defined projects, extending to the year 2000. The project size varies between several hundred thousand dollars and 160 million dollars. It is becoming apparent that the emphasis in the CAM systems is increasingly spreading to the higher level supply systems and on up to management information systems. As a result, the currently installed PPC system with the MRP2 philosophy is increasingly proving to be the weak point as compared with the currency demands of the operative levels. A new system, which is currently at the stage of being selected, should be of help here.

In total 200 employees at LTV are available for the development of the CIM systems. In general, however, only concepts and program specifications are developed; the implementation is entrusted to external software houses or contractors. The foundation of the CIM concept is a 6-level functional model with the individual levels: management, factory, center, system, cell, and equipment. This functional breakdown makes it easier to incorporate the multiplicity of defined projects.

The CIM training of employees up to the management level is accorded considerable prominence.

The efficiency achieved by the flexible machining cell, which has already been in operation for several years, and which has produced a tripling of productivity and the saving of a total of 20 million dollars, appears to ensure the efficiency of the new systems. ROI (return on investment) is calculated for each of the projects within the CIM strategy. Here, intangible (qualitative) factors of influence are also taken into account. The important qualitative factors at LTV are quality and time objectives. Consideration of economic efficiency is a prerequisite for the correct CIM design: "If you cannot identify the benefits, you cannot design the system".

Of course, there has also been unfavorable experience as regards efficiency. The originally isolated use of CAD was supposed to bring about a reduction in the costs of aircraft development. In fact, no cost reduction was achieved, but rather the computer support "merely" led to better design, since more alternatives could be taken into account. At the moment, therefore, closer integration of design and production is paramount, in order to achieve a real reduction in costs.

d. Westinghouse Electrical Corporation, College Station (Texas)

The Factory of the Future built on a green meadow for a capital outlay of 24 million dollars is the CIM show-place of the electronics giant Westinghouse. The 500 employee factory was set up with the strong support of the central departments Computer Integrated Manufacturing Systems and the Technology Center in Baltimore, with the aim of achieving a cost structure that is competitive on a world scale. The plant produces electronic boards, which are delivered to other production facilities within the group, where they are built into electronic systems for aircraft. The assembly system, which is characterized by the use of robots, is capable of building 6,000 different types of boards. At present 300 different types of boards are produced per month. The daily output is 200 boards. In order to achieve the high flexibility of diverse product types a close link between the CAD systems and the production systems has been created. To achieve this, however, considerable computer-technical difficulties had to be overcome, since at the head office in Baltimore the commercial databases are installed on HP systems, the CAD databases on UNIVAC systems and the general host functions are handled by IBM systems. Correspondingly, an IBM 4341 in the Texas works performs a bridging function to the central systems. Below this there is a three-tiered level of HP and DEC computers. It is therefore not surprising that an "intelligent" interface program system had to be developed in order to transform the geometry data from the CAD systems into production-oriented control instructions at the CAM level. It should be noted that not only changes of format are effected here but also various plausibility checks and extensions from the production level are incorporated. Special emphasis is placed on the consideration of production suitability at the development stage (see Fig. D.II.04).

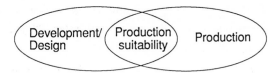

Fig. D.II.04: Production suitability

Not only production control is supported, but also the testing and feedback of production results. An integrated operational data collection system provides data concerning quality, production, costs and workforce employment to the relevant processing systems.

The considerable support from the central head office departments made it possible to set up the system with only twelve in-house systems analysts.

In developing software, user wishes could be taken into account by using prototyping procedures. In total, 160 managers were familiarized with the screen forms during development of the system. The training of employees up to top management is a prerequisite for the effective use of the system.

The performance enhancements and efficiency improvements achieved are illustrated by numerous indicators. Fig. D.II.05,a shows first a comparison of the initial position, or its extrapolation, before the works were founded, Fig. D.II.05,b shows the form of the corresponding curves on the basis of values achieved and expected in the new works.

Given growing product complexity traditional processing reckons with a falling initial yield (i.e. without improvements), the so-called first-time yield. Simultaneously, increasing costs and delivery times (the interval between order receipt and shipping) are also expected.

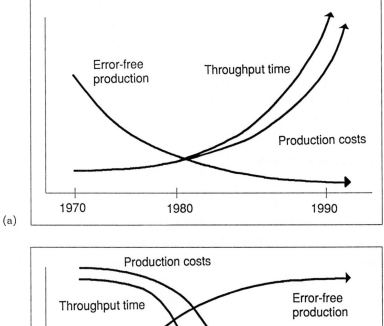

Fig. D.II.05: Trend paths before (a) and after (b) automation

With the help of the CIM technology these trends were transformed into their mirror images. The yield rose from 61% in 1983 to 95% in 1987. The delivery time fell from 12 weeks to 2 weeks. Between 1981 and 1987 costs could be reduced by 55%. By establishing a material acquisition center the procurement time, which in 1981 had averaged 24 weeks, could be reduced to 4 weeks in 1987; availability rose from 47% in 1981 to around 90% in 1987, while at the same time material costs (storage, inspection) fell by 60%.

Automation also altered the cost structure. Direct pay-roll costs were reduced from 50% to 15% of total costs; at the same time indirect costs rose from 15% to 35%. The proportion of material costs is therefore now 50% as compared with 30% before automation.

Further cost reductions which might be achievable from additional CIM projects, therefore, concentrate on material costs.

The efficiency figures are all the more impressive since cost reduction is not automatically reflected in increased profitability of the enterprise. Since the main customer is the Air Force profit is linked on a percentage base to costs. The advantage to Westinghouse is therefore primarily the gain in terms of CIM implementation experience, and in the demonstration character of the Factory of the Future for other customers.

e. DEC Works, Springfield (Massachusetts)

In 1971 the factory was installed in buildings originally constructed as a weapons factory by George Washington in 1795. In the factory disk systems of varying complexity and size are produced. The works have an internal turnover of one billion dollars, which is around 10% of the entire DEC company turnover. The value of the products varies between 1,500 and 500,000 dollars. Of the 700 employees only 25% are manual workers, around 250 employees are concerned with material logistics and 250 with administration and financial tasks. In computing 42 employees are responsible for development and maintenance and the hardware and software areas.

After the takeover in 1971 initially unqualified employees were hired and given a systematic training in high technology. Of course, the high level of automation has reduced the number of employees from 1,500 five years ago to the present 700; employees who were released were given jobs in other DEC works. The high proportion of minorities (60%) and employees from 30 different nations should be given special mention.

For the entire works there exists one CIM strategy, shown in Fig. D.II.06 by the CIM areas production process, manufacturing strategy, production planning and control, product and process technology and computing technology and their associated products and individual projects. All projects are the responsibility of the works. They are justified internally on the basis of return on investment calculations, whereby considerable calculation difficulties can arise for sub-projects; for this reason analysis is increasingly being conducted in larger integration complexes, in order to eliminate overhead blocks. Given the close production interdependences between the DEC works, the DEC-wide MRP2 system MAXCIM is employed.

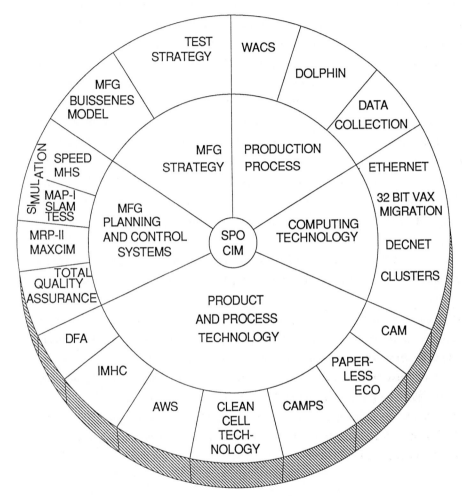

Fig. D.II.06: CIM projects
Source: *DEC*

The essential goals of the CIM strategy are:

- improving customer satisfaction,
- reducing development times,
- increasing product reliability,
- reducing costs.

These goals are brought about by the careful design of the factory itself, of material management and of the information flow between development and production. Suppliers are also included in the CIM model. They are required to use statistical process control systems (SPC) with prescribed test procedures. The necessary programs and PCs are made available to them by DEC. The number of suppliers has, in the process, been reduced from a previous 700 to a current 300 - a further reduction of 100 is considered plausible.

To implement the CIM chain of an integrated engineering database integration tools developed in the DEC-CTC-Center are employed. The dramatic effects of a development chain comprising all the primary data is obvious from the fact that a product amendment (ECO = Engineering Change Order), which previously required 90 days to be implemented in the works, can now be effected by a paperless system within two days. This means that the product changes generated in development are further processed almost in parallel within the PPC system, quality assurance, the storage system and the operational data collection system, and the corresponding planning changes initiated. The old parts numbers are blocked for the human planners until all the changes have been reconciled with each other.

The design is carried out at distant DEC locations - the close informational interdependences are handled by DEC-Net.

Direct pay-roll costs have been reduced from the previous 15% to a current 4% of product costs. In this way the works are capable of competing with cheap labor nations such as Singapore, Korea, etc.

The product costs are calculated at an early stage in the development of the product, by way of a link created between the production and product cost systems. This integration incorporates the principles of design stage cost estimation.

Given the high proportion of material costs in total costs, the stress in coming years will be placed on improving logistics. An initial CIM set-back has already been experienced here, since the introduction of an in-house developed driverless transport system has proved a failure. Improved technology should provide the missing link in the logistic chain between the storage racks and assembly line.

f. Summary

Although the statistics and prominent business magazines make reference to the tardiness of American manufacturing technology, as compared with European and Far East production technology, the examples presented show that America is in the process of catching up on a large scale. This applies not only in the electronics sector but also in mechanical manufacturing. The Government is also participating in this process, particularly by awarding military contracts to high technology industry. The importance of

CIM to American industry is indicated by slogans such as "Texas replies to Japan" (Westinghouse) or "We're making America even greater". In PR films employees are interviewed who report almost euphorically about job enrichment and job enlargement through CIM systems.

The quality of production and the product is increasingly being recognized as the essential goal. Precise recording of quality values achieved at the workplace is not uncommon. They do not even flinch from direct monitoring of employees by other employees. For example, processed parts are passed on directly from one workplace to the next, so that an employee can be admonished immediately by his neighbor for poor or incomplete work. The mixture of high technology and the typical American love of practical rules is obvious in all other situations, physical space will not be left free within an assembly line in order to avoid the holding of stocks, even though this would have been available: Where there is no space you cannot store any parts, or alternatively: free space generates unnecessary stocks.

The short term economic analysis, which has been dominant in America for a long time as a result of the short term three-monthly stock exchange reports (quarterly reports), is gradually being extended or even replaced by strategic considerations.

International comparisons are made in order to identify the weak points of their own procedures as compared with the competition.

It is characteristic that the leading CIM industrial enterprises have understood the wider effects of the integration principle beyond simple production automation through to all the functional areas. CIM is increasingly being equated with CIE (Computer Integrated Enterprise).

E. CIM Promotion Measures

I. The CIM-Technology-Transfer-Centers' Program Using the Example of the Saarbrücken Site

(Dipl.-Ing. Thomas Geib, Dipl.-Ing. Arnold Traut, Institut für Wirtschaftsinformatik (IWi), University of Saarland)

a. Tasks of the CIM-Technology-Transfer-Centers

German Ministry for Research and Technology founded a program called "Production Technology 1988 - 1992" to promote the implementation of CIM solutions in Germany. In the scope of this program, industrial enterprises - mainly those from mechanical engineering, plant engineering and construction - could obtain financial support for creating and implementing CIM concepts. A further measure, which was financed by this program, is the so-called "wide efficacious diffusion via CIM-Technology-Transfer". This part of the promotion program supported the building-up of CIM-Technology-Transfer-Centers (CIM-TTC) at 21 university departments in Germany (see Fig. E.I.01). CIM-TTC Saarbrücken being part of Institut für Wirtschaftsinformatik (IWi) at the University of Saarland is affiliated to the only business administration institute in the project. All the other CIM-TT partners are affiliated to institutes of the engineering science area. The cooperation among the CIM-TTCs was handled by BMFT's department "production techniques", which is located at Kernforschungszentrum Karlsruhe (KfK). Five of the CIM-TTCs had been installed in the new Eastern German states after the unification in 1990. Their sponsorship was prolonged until 1995, while the program ended for the 16 CIM-TTCs in the West But most of those CIM-TTCs exist further, e.g. CIM-TTC Saarbrücken pays its way by seminars and research projects which are hold in cooperation with Institut für Wirtschaftsinformatik.

In addition to financial support, the intention of the CIM-TTCs is to provide conceptual help to those enterprises that are interested in CIM-technologies. The CIM know-how of the research institutes involved in the project is made available on a broad scale. The target group, at which the transfer of technology is aimed, are small to medium sized enterprises. On the one hand those enterprises usually have good prerequisites for lean enterprise structures and innovative production (e.g. considerable organizational flexibility or the use of standard software packages). On the other hand they lack the necessary

252

Hamburg
Technische Universität
Hamburg-Harburg
Arb.-Bereich Fertigungstechnik I

Bremen
Universität Bremen
Bremer Institut für Betriebstechnik und
angewandte Arbeitswissenschaft

Dortmund
Universität Dortmund
Institut für spanende
Fertigung

Bochum
Ruhr-Universität Bochum
Lehrstuhl für Produktionssysteme
und Prozeßleittechnik

Aachen
RWTH Aachen
Werkzeugmaschinenlabor

Darmstadt
Technische Hochschule Darmstadt
Institut für Produktionstechnik
und spanende Werkzeugmaschinen

Saarbrücken
Universität des Saarlandes
Institut für Wirtschaftsinformatik

Kaiserslautern
Universität Kaiserslautern
Lehrstuhl für Fertigungtechnik und
Betriebsorganisation

Karlsruhe
Universität Karlsruhe
Lehrstuhl und Institut für Werkzeug-
maschinen und Betriebstechnik

Kiel
Fachhochschule Kiel
Institut für
CAD/CAM-Anwendung

Wismar
Technische
Hochschule Wismar
Fakultät Maschinenbau

Stuttgart
Universität Stuttgart
Institutsverbund
Fertigungstechnik

Hannover
Technische Universität Hannover
Institut für Fertigungstechnik und
spanende Werkzeugmaschinen

Braunschweig
Techn. Universität Braunschweig
Institut für Werkzeugmaschinen
und Fertigungstechnik

Berlin
Technische Universität Berlin
Institut für Werkzeugmaschinen
und Fertigungstechnik

Magdeburg
Technische Universität
"Otto von Guericke"
CIM-Technologie-Transfer-Zentrum

Dresden
Technische Universität Dresden
Institutsverbund Fertigungstechnik
und Werkzeugmaschinen

Chemnitz
Technische Universität Chemnitz
Fachbereich Maschinenbau II

Suhl
Technische Hochschule Illmenau
Institut für Präzisionstechnik
und Automation

Erlangen
Universität Erlangen-Nürnberg
Lehrstuhl für Fertigungsautomati-
sierung und Produktionssystematik

München
Technische Universität München
Lehrstuhl für Werkzeugmaschinen
und Betriebswissenschaften

Fig. E.I.01: CIM-Technology-Transfer-Center Sites in Germany

specialist knowledge in estimating the prospects as well as risks of CIM, and they usually
don't have a survey of the steps of planning and introducing CIM.

The operational spectrum of the CIM-TTCs is wide-ranged:

❑ organization and hosting seminars on subjects in the area of computer integrated
production,

❑ organization of meetings for the exchange of CIM knowledge,

❑ presentation of examples about CIM implementations by demonstrating CIM
operations, which have been developed at each site.

CIM-TT cross-section topics	Aachen	Berlin	Bochum	Braunschweig	Bremen	Darmstadt	Dortmund	Erlangen	Hamburg	Hannover	Kaiserslautern	Karlsruhe	Kiel	München	Saarbrücken	Stuttgart
1. CIM definitions and CIM basic elements	x		x		x		C									
2. CIM strategy as a part of the enterprise strategy	x					x				x					C	
3. Analysis and restructuring of the factory		x			x	x	x		x	C						x
4. CIM planning and introduction strategy	x						C					x			x	
5. Personnel development and qualifications	x	x	x	x												C[1]
6. Networks, communication technology	x	x	x	x	x				x	x[2]	x					C
7. Interfaces	x		x	x					x				x	C		
8. Databases for CIM	x	C	x								x				x	
9. Simulation in CIM	C		x			x	x	x				x				
10. CAD/CAM-centred linkage of CIM elements		x				x		x	x		x			C	x	x
11. PPS-centred linkage of CIM elements	x	x		x	x			C						x	x	x
12. CAQ-centred linkage of CIM elements				C	x			x			x					x
13. CIM production islands			x	C	x		x					x	x	x	x	
14. Assembly planning in CIM	x	x						C				x			x	x
15. CIM in one-off production and assembly				x	x	C										
16. Expert systems in CIM		x							x	x	C			x		
17. Workshop information systems in CIM		x	x											C	x	

x = Involvement   C = Coordination   1 Prof. Bullinger   2 Prof. Rall

Fig. E.I.02: CIM-TT cross-section topics

b. Seminars on CIM

In the project's starting phase, the CIM-TT partners - in cooperation with the project sponsor - defined 17 so-called "cross-section topics", which comprehensively cover the range of issues referring to "CIM" (see Fig. E.I.02). These topics were developed collectively by the CIM-TT partners. In this way, a large amount of training material was worked out rather quickly.

In order to construct these contributions on a basis as broad as possible, external university institutes, organizations and institutions as well as employees of industrial firms were involved in the process. The use of uniform hard- and software allows paper and diskette versions of all topics to be obtained at all sites. Using this collection of material, individualized collections of papers can be compiled for the seminar meetings according to the specific participating groups. The target groups at whom the range of seminars are aimed are mostly members of the company management and works councils, members of the intermediate technical and business management levels (engineers, foremen, specialists).

In the scope of the collective production of training material, CIM-TTC Saarbrücken played the leading role in preparing the cross-section topic "CIM Strategy as Part of the Enterprise Strategy". Here, CIM is interpreted not merely from the point of view of the technical integration of computer systems, but as an enterprise's essential strategic decision, which can only be implemented in medium to long terms. Moreover such a basic decision has decisive influence on the competitiveness of the enterprise. In addition to suggested strategies for developing an enterprise-wide CIM concept, information about the technical integration of CIM components is given. The business-organizational and the personnel implications of introducing CIM are further important aspects. Practical examples of CIM implementations complete the presentation.

The seminar program offered at CIM-TTC Saarbrücken not only covers the contents of its own cross-section topic. Of course, meetings are envisaged on all relevant CIM topics. In all seminars not only the technical issues of computer integrated production are considered, but also the business-organizational aspects are highlighted.

Based on the cross-section subjects, a book series was published in 1992/93, which gives a qualified survey on each aspect, when introducing CIM. The series is called "CIM-Fachmann" ("CIM expert"). Volume 2 describes IWi's cross section topic "CIM Strategy as Part of the Enterprise Strategy" and is published under this title.

c. Demonstration Possibilities

In order to support and illustrate the seminar meetings, CIM-TTC Saarbrücken has developed a CIM demonstration plant. This prototype is an practical example for an integrated order transaction in an enterprise: order acceptance - construction - master planning - material and capacity management - work planning and scheduling - several production stages - final delivery. Systems used in different industries, such as CAD/CAM, PPC-systems, production-Leitstand, drilling and milling center as well as industrial robots, are installed. All systems were commonly integrated into one information and material flow system. This plant demonstrates how business processes pass and how these business processes are accompanied by the information flow.

The demonstration plant is based on Scheer's Y-CIM-model and covers the planning and dispositive level, the control level and the operative level. It's configuration is given in Fig. E.I.03.

The used computing systems include IBM AS 400, IBM 6150, HP 3000, HP 9000, DEC VAX-Station 3100, IBM IC 7562, Engel&Stiefvater ES68, IBM PS/2-Systems, HP-Vectra-PC-Systems, IBM-compatible PCs. The computer systems are networked by a Ethernet 802.3 and an IBM token ring network. Different protocols can be used. The use of remote data links enables the access of the LAN of IWi located on the university campus as well as the computer center of the University of Saarland.

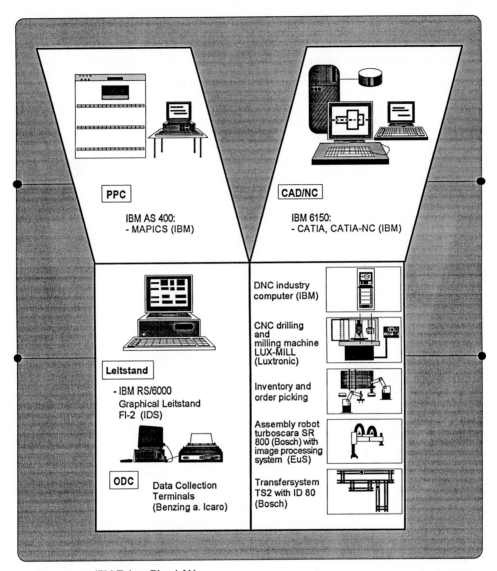

IBM-Token-Ring-LAN
Ethernet

Fig. E.I.03: Configuration outline of the CIM demonstration plant at CIM-TTC
　　　　　Saarbrücken

The available software systems cover almost the entire functional scope within CIM. For production planning and control, MAPICS from IBM is in use. CAD constructions can be drawn up on CATIA from IBM.

The graphical Leitstand FI-2 from IDS Prof. Scheer GmbH is used to support job-shop control. The transformation of design drawings into NC programs is handled by the systems PRO*NC and CATIA-NC. Diverse database, word processing, desk-top publishing, simulation, animation and drawing systems are also at hand.

The shop-floor, consisting of processing center, production line and assembly cell, comprises the components:
- LUXTRONIC CNC-drilling and milling processing center LUX-MILL,
- two LUX-ROBOT industrial robots,
- BOSCH turboscara SR 800 assembly robot, including a robot control RS/82,
- Fischertechnik racking system,
- the data collection system Icaro and three BENZING terminals Bedem 954,
- Bosch transfer system TS 2, including a identification- and datamemory system ID 80,
- Bosch PLC CL 300,
- Engel&Stiefvater image processing system.

This configuration makes it possible to represent the order transition through a CIM-oriented enterprise. In this enterprise quartz desk clocks are produced. The basic body of these clocks is an aluminum block, on which, in one step, a monogram is engraved according to requirements and recesses are milled to house the mechanism and various fitted parts (see Fig. E.I.04).A "customer" can choose among ten variants, which differ in the parts to be inserted in the basic body.

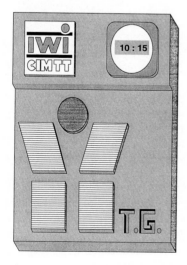

Fig. E.I.04: CIM product - "quartz desk clock"

The production starting with the order works this way:

The order specifies the type and number of fitted parts, the desired monogram, the number of clocks to be produced and the desired delivery date. The possibilities for influencing events open to the "customer" concerning the structure of his order in the demonstration enterprise demand flexibility with respect to the production and subsequent use of the geometry data in manufacture and assembly, as well as with respect to the scheduling of the number and kind of parts to be assembled. This flexibility is achieved by linking the CIM modules used.

Order transaction begins with the input of the order into the PPC system, which confirms the order after a successful availability check.

The ordered clock is designed with the help of a CAD system. The NC programs for the drilling and milling process are created in a NC-programming module of the CAD system. The bill of materials which is also produced by the CAD system is passed to the PPC system. There the orders are scheduled according to deadline and capacity. The stock of orders hold in the PPC system are made available to the graphical Leitstand FI-2 for a detailed scheduling and machine loading. The availability checks are carried out and the order is released for production. The NC programs, which are needed for processing the raw material at the drilling and milling processing center, are transmitted from the CAD system to the DNC system via LAN. On request they are forwarded to the CNC machine control of the drilling and milling process center.

After the mechanical processing, the basic clock body is positioned on a workpiece carrier of the transfer system, which connects the three stations of the following production line:

- Input of order number,
- Inventory and order picking,
- Assembly.

A general information and material flow along the transfer system is guaranteed by the used computers, automation systems and production devices. Most important part of the information flow is represented by an identification and data memory system, which is installed at each station. It consists of locally fixed read-and-write stations and mobile data carriers which are fastened on the workpiece carriers. Communication between read-and-write station and mobile data carrier is controlled either by a host computer connected to the read-and-write station or by a program that is kept in the read-and-write station.

At the station "input of order" the order number is transmitted to the mobile data carrier. This order number contains the variant number. After this procedure the workpiece carrier is released so that it is driven to the station "inventory and order picking". This cell contains two robots connected by a short material flow band with a circular assembly table and a racking storage system. The fitted parts, that are needed for each order, are directly consigned to the workpiece carrier alongside the basic body. The successful consignment is written into the mobile data carrier.

Prepared in this way, the workpiece carrier is being transported along the transfer line to the assembly cell, where the data in the mobile data carrier are analyzed in the first step. The assembly will only take place, if the data confirm a successful part consignment in the station before. The robot control reads the variant number from the mobile data carrier. A vision system, connected to the PLC, catches the current image of the workpiece carrier. The coordinates of the needed parts are calculated in the image processing system and transmitted to the PLC, where the robot arm's movement-trajectory of the following assembly is created. After the assembly is finished, the workpiece carrier is released and moved to the final stop, where the ordered quartz desk clock is taken off the working carrier to be given to distribution.

Feedback concerning the progress of the order is realized at the processing center and the assembly station by using terminals of a data collection system. The recorded data are first put into intermediate store in the data collection computer, pre-processed and passed via the control center to the PPC system. Here the delivery notes and invoices are generated on order completion.

The PPC system supports the entire order transaction: order acceptance - store handling - rough time scheduling - invoice writing. The tasks of fine scheduling are performed by a Leitstand system, which allocates the orders to the device resources. The staff being responsible for planning can do this is in dialogue with the Leitstand's electronic planning board.

In the scope of current IWi-projects, the model factory is being extended permanently, so that it is kept on state of the art. E.g.:

Within the European ESPRIT II project No. 2527 "CIDAM", a universal usable tool for managing data-base-interfaces was developed. This INterface MAnagement System (INMAS) initiates the data exchange between applications by receiving trigger-signals. If any data change in the data base of one involved production system, this system sends a trigger to INMAS. INMAS analyzes the trigger and selects an activity chain referring to the trigger. In order to update changed data, this activity chain contains data and transformation rules of those target systems that keep the same data redundantly. The demonstration plant of CIM-TTC Saarbrücken was an ideal testbed for INMAS and it is going to be expanded by means of multimedia at the moment.

II. CIM in Central and Eastern European Countries

(Dr. Thaddäus Elsner, Institut für Wirtschaftsinformatik (IWi), University of the Saarland, Saarbrücken)

a. Introduction

After the collapse of communism, Central and Eastern European countries are in the middle of important structural changes. Articles in journals and magazines trend to focus on politics, society and the economy. However, this paper discusses not only transformations in socio-economics, but also in research and development and higher education in Central and Eastern European countries.

1. Universities

In the past, big universities in Central and Eastern Europe (see Fig. E.II.01) had relatively good access to Western literature. As a result, they were well informed about the state of basic research in Western Europe. Since the break down of the wall, universities have had many problems that have lead to a loss of image. For example, academic teaching staff earn very little and officials of the old regime remain in their positions. Consequently, it is very difficult to get young academic talent which prevents a younger generation from breaking down the old personnel structures. In addition to personnel problems, the adaptation of higher education to the Western European level has been prevented by other

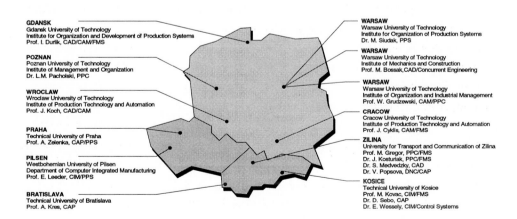

Fig. E.II.01: Research and education centres for CIM in Poland, Czech and the Slovak Republic

facts such as the technically obsolete equipment of the Departments for Computer Integrated Manufacturing which did not refer to the demand of industry and consequently were obstacles to introducing modern practical concepts within higher education.

However, the answer for research and higher education in Central and Eastern European countries is not simply to import western ready-made solutions. Only cooperation and the adaptation of concepts to the specific needs of these countries will allow the rehabilitation of research and development, higher education and pratical application in Central and Eastern Europe.

2. The Economy

Following the collapse of communism, companies in Central and Eastern European countries were abruptly confronted with free market rules. Today, companies face competition that simply did not exist in a planned economy, as well as the loss of markets as eastern trade breaks down and former customers declare their insolvency. Furthermore, they are having great difficulty accessing new markets in the European Union. The reasons are lack of competitive products, quality problems, obsolete manufacturing technology and non-qualified staff. On the contrary, West European companies are gaining in eastern markets which increases competition even more.

In addition to the changes that must be made to a market economy, privatisation in most Central and Eastern European countries is happening too slowly, especially for former large national companies. The adminstrative ineffectiveness is amplified by traditional business customs such as in Poland where big companies only allow minor participation of outsiders for fear of loosing control. Consequently, foreign, financially powerful companies are not interested in investing due to the fear of ineffective management and capital loss. In the Czech Republic foreigners are not allowed to aquire ground which is a politically motivated rule with origins in the past. Due to the unresolved situation of property and ground in many Central and Eastern European Countries, the readiness of industry to invest is not sufficient.

b. Cooperation with the European Union

Poland, Czech and the Slovac Republic will probably become members of the European Union in the near future. Furthermore, Central and Eastern European Countries and the former States of the Soviet Union will be important future markets for the European Union. In addition to business aims, an important political goal for the European Union will be the economic integration of Central and Eastern European countries and the former States of the Soviet Union in order to maintain stability in the biggest "economic zone" of the world.

1. Cooperation in "Higher Education", the TEMPUS program

TEMPUS (Trans-European cooperation scheme for higher education) passed the Council of Ministers of the European Union in May 1990. In 1993, it was renewed for four more years. The TEMPUS program is part of a large scale program to reform the economies and societies of Central and Eastern European countries (Tempus-PHARE) as well as the former states of the Soviet Union (Tempus-TACIS). One of the priority sectors within the TEMPUS program is the development of human ressources in order to reach West European levels of business administration education.

Within the scope of the TEMPUS program, the Institut für Wirtschaftsinformatik (IWi) at Saarland University (Germany) elaborated the UNIVERS project (University-Cooperation Business Information Systems) which is financed by the Commission of the European Union (Joint European Project JEP-04221). Partners within the UNIVERS project of the Institut für Wirtschaftsinformatik (IWi) are Warsaw University (Poland), three Polish companies and the Institut Commercial de Nancy (France).

The three year project (September 1992 - August 1995) aims to reform business administration studies in Poland by introducing a new curricula in business administration emphasizing business information systems. Present studies in business administration in Europe have been critically examined, and new curricula have been developed. Graduates holding a degree in business information sciences are mainly involved with the development and introduction of information systems and the elaboration and introduction of new forms of organization. The Western European experience shows that the solution to such problems is a prerequisite to competitiveness. In a first step, the course will be introduced at Warsaw University which will be the pilot University. Later the dissemination of the new knowledge will spread to all faculties of economic science at other Polish universities which will then be able to educate specialists who can work on the problems of transitioning Poland to a market economy.

The UNIVERS-Project includes the following activities:

❏ Development of curricula and teaching material for the new course of studies: business administration emphasizing business information sciences

❏ For the business information sciences subject: development of teaching materials for lectures, seminars and laboratory sessions (in Polish); hardware and software for laboratory sessions at a Production Planning and Control System (PPC) and a PPC-Simulator

❏ Structural development: computer laboratories, provision of new literature and of modern equipment for the library in Warsaw

❏ Study abroad program for students from Warsaw in Saarbrücken and Nancy (language classes, studies, practical training)

❏ Study abroad program for teaching staff from Warsaw in Saarbrücken and Nancy (studies of curricula, further education)

❏ Teaching of business information sciences at the University of Warsaw by lecturers from Saarbrücken and Nancy

Education within the UNIVERS program of the Institut für Wirtschaftsinformatik (IWi) starts in the third project year (September 1994). Selected lectures are already being offered at Warsaw University. In addition, students who take part in the study abroad program are included in setting up the project. It is assumed, that at the end of the cooperation Warsaw University will be able to continue the new education for students as well as for industry and commerce.

2. Cooperation in "Research and Development", the COPERNICUS program

Besides programs for higher education, the European Union supports joint research projects across Europe to develop and enhance existing scientific expertise and to promote technology transfer to the mutual benefit of the Central and Eastern European and European Community partners. The main goal of the cooperation in science and technology with the Central and Eastern European Countries (COPERNICUS) is the strengthening of their research capacity and reorientation of their socio-economic needs which are of prime importance for the successful transformation of their economies. The COPERNICUS-Proposal of 1994 is the third to have taken place under the Cooperation in Science and Technology with Central and Eastern European countries. The first proposal was issued in May 1992. The second in 1993 was limited to participation of Central and Eastern European countries in ongoing projects within five specific sections (Environment, Non-nuclear Energy, Nuclear Fission safety, Biomedicine and Health, Human Capital and Mobility). The COPERNCUS-Program from 1994 will focus Research and Development in the following sectors as a high impact on the development of the economic and social systems is expected:

❏ Information Technology
❏ Communication Technologies, Telematics and Language Engineering
❏ Manufacturing, Production, Processing and Materials
❏ Measurement and Testing
❏ Agro- and Food Industries
❏ Biotechnology

Computer Integrated Manufacturing and Engineering is a priority area within Information Technology. It includes :

☐ the development of advanced IT solutions for more efficient industrial operations and processes
☐ integrating engineering, logistics, operations, process automation and business functions so that social, organisational, economical and enviromental needs are taken into account.
☐ architecture and infrastructure for Computer Integrated Manufacturing and Engineering,
☐ management and design of industrial enterprises,
☐ mechatronics, robotics and sensing technologies,
☐ microsystems manufacturing and integration.

c. Summary

In conclusion, the ultimate goal of European cooperation is to not only close the gap between Eastern and Western Europe but also to generally promote a move within academia from a too theorical emphasis to one that is more practical (for example Production Planning and Control). Another objective is to highten the awareness of industry to research relevant to their commercial plans. Therefore, it is essential to create a framework for attaining research and development targets via active larger scale multinational cooperations. This would increase scientific efficiency as well as promote its efficient transfer into practical application which remain the objectives of any modern economy. In this context, strengthening the relation between industry, research organisations and universities is an important consideration. Strong preference is therefore given to projects in applied research directly concerning countries of Central and Eastern Europe which include industrial enterprises or small and medium-sized companies. Supporting such joint research projects is intended to provide a platform for industry, research institutes and universities to cooperate and coodinate innovative research and development and educational activities, in a given field, to the mutual benefit of the countries involved.

III. Cooperation Model of CIM Technology Development and Transfer to Brazil

(Prof. Dr. August-Wilhelm Scheer, Dipl.-Ing. Arnold Traut, Dipl.-Kfm. Markus Nüttgens, Institut für Wirtschaftsinformatik (IWi), University of Saarland, Prof. Dr. Heitor Mansur Caulliraux, Prof. Dr. Roger Boyd Walker, Eng. Arnaldo Ferreira Sima, Coordenação dos Programas de Pós-Graduação em Engenharia (COPPE), Federal University of Rio de Janeiro (UFRJ))

a. CIM potential in Brazil

Brazil is supposed to be the most important industrial nation in Southamerica, because it succeeded in building-up automotive industry, mechanical engineering and high-tech enterprises in the last decades. In this context, however, one mostly disregards that these industries developed in the shadow of restrictive import limitations and protective duties. E.g., up to 1991 there existed a nearly complete import prohibition on cars and computers. This means that exporters to Brazil have merely two choices: Either they construct their own production plant in Brazil or they cooperate with domestic partner enterprises. The lacking international competition caused a considerable increase of prices. A machine compulsorily bought in Brazil usually costs double the world market price. If one furthermore regards the missing discipline in planning, which causes increased dispositive and stock costs, it is obvious that the automotive plants in Brazil complain of much higher cost structures than their European mother enterprise. A liberalizing of imports by the government in 1991 abolished import limitations and reduced the duties to a certain degree, but it caused immense problems which arise with stronger competition. Brazil passes in the middle of this decade an important stage of reorganization of its industrial landscape. To stand this, the Brazilian economy needs to reach world level in industrial production.

Bottleneck number one is the lacking financing. Nowadays' high inflation rates (up to 40 % per month), the saving politics of the government and lacking support programs in research prevent the installation of necessary production systems and computer devices. The uncertainty in planning still effects large stocks, because security aspects dominate.

The use of CIM technologies can be an effective measure to become competitive. Moreover, product quality can be improved. The essential CIM-potential in Brazil will be found in human resources, e.g. the high efforts of younger professors at the universities of São Paulo, São Carlos, Florianopolis, Rio de Janeiro and other universities in the south of Brazil are promising starting points. Many of them studied in Europe, especially in Germany, and within their graduation they got their CIM knowledge. Cooperations with European institutes can keep them on state of the art. At the universities they skill very good engineers having well-founded know-how. A detailed analysis is given in *Scheer, CIM in Brasilien 1991*.

Examples of realized and promising interdisciplinary CIM research projects at universities are

- ❏ UFRJ/SENAI/IWi - Laboratório do Y-CIM-Lab), CETEMM Euvaldo Lodi, Rio de Janeiro/RJ,
- ❏ USP - Laboratório de Máquinas Ferramentas (LAMAFE), São Carlos/SP,
- ❏ SABO - Sistema de Chão de Fabrica (SFC), São Paulo/SP.

In the following part, the building-up of the Y-CIM-Lab is presented as a successful example of an international cooperation model. CIM-COOP is a strategic cooperation project between the Brazilian partners Federal University of Rio de Janeiro (UFRJ) and SENAI, Regional Department of Rio de Janeiro, as well as the German partner Institut für Wirtschaftsinformatik (IWi).

b. CIM-COOP: Example of an International Cooperation Model

The objective of the CIM-COOP project is to develop a model for a CIM-Technology cooperation between German and Brazilian partners, including transfer and bilateral development of technologies. It deals with the process of economical rearrangement and adaptation of small and medium-sized enterprises (SME) and includes different steps of activities coordinated by national and international partners. The cooperation is focused on planning and adjusting modern infrastructure of primary and continuing skill for employees using integrated technology for production. Its main goal is the training of employees in order to enable them to implement the concept of integrated technology for production. Thus it is intended to build-up a modern and competitive industry in Brazil focusing on small and medium-sized enterprises.

1. CIM-COOP Consortium

The compound of CIM-COOP was selected in order to guarantee a multiplication effect of all measures. Several kinds of partners are involved in the project CIM-COOP. On the one hand, two scientific and research institutes, a Brazilian and a German one, as well as Brazil's largest apprenticeship organization form the nucleus partners of the CIM-COOP-Consortium:

❑ IWi - Institut für Wirtschaftsinformatik (IWi) at University of Saarland, Saarbrücken/Germany,

❑ COPPE/UFRJ - Coordenação dos Programas de Pós-Graduação em Engenharia (COPPE) at the Federal University of Rio de Janeiro/Brazil (UFRJ), especially the Integrated Production Group (GPI),

❑ SENAI - Serviço National de Aprendizagem Industrial, Regional Department of the state of Rio de Janeiro/Brazil.

On the other hand, there are numerous German and Brazilian companies involved. While the institutes develop concepts and thus ensure the scientific basis of the project, SENAI skills young people mainly in technical jobs. A close relation to industry is guaranteed by the connecting links between the nucleus partners and industrial companies. IWi as well as COPPE base these relationships on many cooperations within projects. The German company-group of CIM-COOP formed in the scope of a Hannover-fair's activity in cooperation with IWi, as the realization of the Y-CIM model into a demonstration plant was successfully shown. They are interested in cooperating in CIM-COOP in order to build-up and intensify their contact to Brazil. Some of them already are represented in Brazil. The companies directly or indirectly involved in the CIM-COOP project and their connection to the nucleus partners are enumerated in Fig. E.III.01.

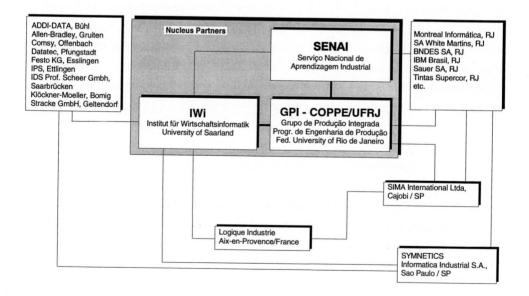

Fig. E.III.01: CIM-COOP Consortium: nucleus partners and involved companies

The national apprenticeship service SENAI is directly involved in Brazil's industry, because it represents the industrial companies' organization and is funded by them. The 27 regional departments are directly responsible for putting teaching programs into practice. SENAI works with 600 operational units of its own, such as Technology Centers, Technical Schools, Professional Training Centers, Training Agencies, Operational Training and Mobile Centers. The regional departments act in close cooperation with industries in their regions, seeking to fulfill demands for skilled labor in accordance with local requirements. These units will be used to offer courses and seminars. In Brazil, there is a strong tendency of the apprenticeship system shifting from federal hands into private charge, and SENAI is the most important private organization. SENAI and the involved companies ensure the realization of developed concepts into practice.

2. CIM-COOP Strategy

Today existing training concepts for SMEs lack an offer for the target group of decision makers. An enterprise's management has the main task to define medium and long term goals and make the decisions to reach them. Beside the technical and financial aspects, the role of training and guiding employees is increasing more and more. These tasks are part of strategy aspects and are called decision-knowledge and employee guidance. As industrial practice shows, the target group's acceptance of computer aided production technologies is a basic condition for an increasing EDP-use in enterprises. In order to have an overview on computer aided technologies, a decision maker has to have compressed basic and special knowledge. Therefore, he should not be taught in detailed knowledge, but in strategic knowledge in order to be able to calculate the consequences of possible decisions.

Such a concept, which is especially directed to qualifying special target groups when introducing and applying CIM components in small and medium-sized enterprises, was developed at IWi. This qualification model follows strictly the integrative idea of CIM. As an open-structured frame concept, it can be used as an aid to develop concrete adapted qualification measures. The seminars for decision makers are divided in three sections becoming more detailed from top to down, as the "training pyramid" in Fig. E.III.02 elucidates. Based on the qualification model, concrete qualification offers such as trainings in the scope of courses and seminars are deducted. In order to ensure a well analyzed and detailed training concept, mainly timestabil training materials are composed, which are supplemented by topical information.

An international project aiming to transfer CIM technology always has to regard the basic qualification concept, which has to be adapted to local circumstances in a first step. Very

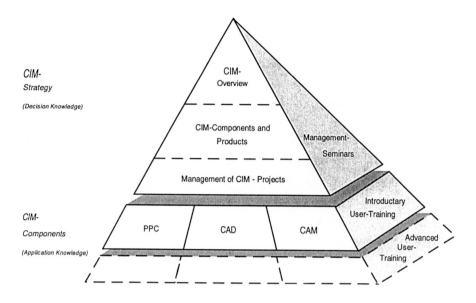

Fig. E.III.02: Training pyramid
source: *Nüttgens, M.; Scheer, A.-W.: CIM-Qualifizierungskonzept für KMU 1991*

important is the exchange of personnel, so that the objectives and problems can be discussed directly. Up to now, mainly the exchange on university level was supported in the scope of pure research projects. Technology transfer requires the exchange of personnel applying the technology and imparting the knowledge to others in order to obtain multiplier effects.

In the case of the CIM-COOP project, following steps are gone:

❑ Exchange of multipliers:

There is an intensive exchange of scientists of all levels. The duration longs from three weeks to six months. The Brazilian groups visiting Germany stay most of the time at CIM-TTC Saarbrücken, but numerous visits to other German CIM-TTCs, institutes and exemplary companies are organized. E.g., in 1993 five Brazilian post-graduate students had a six months stay at IWi in Saarbrücken in order to get theoretically skilled in CIM technologies and concepts focused on the Y-CIM model for industrial integration. Practically the team was trained on a Y-CIM prototype plant, which demonstrates the highest level of automation. Beside many visits, they took part in several industrial courses and thus some promising contacts were established. Another stay of Brazilian post-graduates at CIM-TTC Saarbrücken qualifies them in production management methods and in using software tools based on the ARIS architecture (see *Scheer, A.-W.: ARIS 1992*).

❑ Qualifying decision makers and management staff:
The post-graduates additionally skilled in Germany hold courses and seminars at the university and at SENAI's schools in the state of Rio de Janeiro. For the skilled persons, it is very important that they can immediately be trained on the use of real CIM components, in order to get practical demonstration of the theoretical skill before. The exemplary demonstration facilities being available in the Y-CIM-Lab at CETEMM Euvaldo Lodi, SENAI's largest training center in Rio de Janeiro, are presented in the next point.

❑ Consulting companies in the state of Rio de Janeiro:
Companies of the state of Rio de Janeiro that are interested in integrating components get consultation by the partners of Euvaldo Lodi's Y-CIM-Lab. This consultation covers the conceptual methods as well as the technical methods up to questions of implementing the prepared methods. Current business processes are modeled by using the ARIS architecture. They are analyzed and subsequently optimized.

❑ Diffusion of CIM concepts and technologies in Brazil:
In order to expand technology transfer to the other locations in Brazil, the qualification concepts are adapted to the local circumstances. This step will be decisively supported by SENAI, because it is the ideal partner with its country-wide, regional and local structure.

❑ Development of technical training methods and integrated quality systems by using newest technologies:
In order to support the diffusion of CIM, self-training software systems will be a useful mean. The nucleus partners IWi, COPPE and SENAI develop those self-training systems, especially using multi media technologies. Quality is a decisive criterion of being competitive in the market and has to be integrated in the production process. New technologies can support to guarantee quality, if they are integrated in business processes. Combinations are also imaginable, e.g. in the scope of one step of CIM-COOP project is planned to integrate the quality check images of the Y-CIM-Lab into multi media applications, where they are worked up in training systems. The statistics of the evaluated images give conclusion on the produced quality.

3. CIM-COOP Demonstration Facilities

Beside the theoretical skill, the practical use of devices plays an important role. To enable the Brazilian partners to demonstrate different levels of automation, an integrated Y-CIM prototype is transferred from Saarbrücken to Brazil. It is installed in the Y-CIM-Lab at CETEMM Euvaldo Lodi, SENAI's largest training center in Rio de Janeiro. This production-line represents the highest level of automation beside the semi- and non-automated production devices already installed, e.g. CNC drilling and milling machines, PLC controlled robots, handwork places. Thus, combinations of different components can be demonstrated, and real existing configurations can be indicated by referring to a company's conditions.

The Y-CIM prototype had been developed within a project of IWi in cooperation with ten different industrial companies. It strictly bases on the Y-CIM model. The idea is to manage a computer aided information transaction, including all stages of production. The integrated order processing is demonstrated, considering order entry, construction, work scheduling, manufacturing, quality control and finally shipping.

Exemplary, this CIM demonstration model produces buttons consisting of four separate parts which are composed. Flexibility is demonstrated by the choice of the basic color and an individual name which is printed on top. The CAM-components of the Y-CIM prototype were transferred to Rio de Janeiro and re-installed in the Y-CIM-Lab at CETEMM Euvaldo Lodi (see Fig. E.III.03).

The Y-CIM prototype's concept realizes the integration of computers and control systems as well as varied CIM-application components based on common networks and standardized interfaces. The main item of the model fabric is represented by a production line, consisting of five stations and a connecting conveyance system. Four different Programmable Logic Controllers (PLC) steer the manufacturing processes in compliance with the data of order, construction, parts list and working schedule that are delivered by systems of other companies.

All actions are coordinated and monitored by a process monitoring system, which gives the order to the conveyance system in order to refer it to a palette being not in use at this moment. The order contains the chosen button color and the individual name which will be printed on it. The first station is an inventory station, which contains the single parts that are necessary for the production of a button. After picking the single parts and putting them on a transport palette, they are conveyed to the assembly station where all pieces are formed to a complete button. Subsequently, the ready button is brought on the same palette to the inscription station where an automatic arm puts the button on a turnable table. Using a camera and a low level image processing system, here the button is turned in a right position so that the individual name can be printed in a writing field. The button is moved

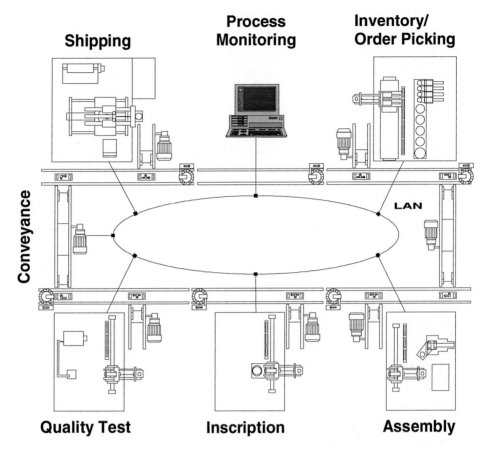

Fig. E.III.03: Concept of the Y-CIM prototype installed in the Y-CIM-Lab at CETEMM Euvaldo Lodi, Rio de Janeiro

Fig. E.III.04: The product of the Y-CIM prototype

along below a print head. When finished, it is put on the palette and is conveyed to the quality station. Here the image of the button is captured by a camera and analyzed in an image processing system which decides, if the name is written correctly in the writing field. Fig. E.III.04 shows the layout of the button. Catching such an image, the image processing system makes conclusions on the produced quality. With a positive result the button is given to the shipping station, else it is rejected at this station.

In the informationtechnical view, the Y-CIM prototype "button plant" represents the control level and the operational level. In a next project step, the production Leitstand FI-2 will be installed, which will control several working cells. Flexible planning strategies enable FI-2 to regard the needs of different productions. Aims like the reduction of transaction times or the improvement of equipment utilization times can be realized easily. It calculates the time scheduling of the orders and their operations at each machine and each working place, including non-automated handworking places. FI-2 is described in detail in chapter C.III.a. To schedule the entire resources (personnel, machines, tools, material) the resource Leitstand RI-2 will be installed too and support the logistic planning. The Y-CIM prototype will be one of these controlled cells, but the one with the highest level of automation integrating diverse systems. The other cells are CNC Labs containing automation machines. The planning level will be completed by the integration of PPC and CAD systems. Fig. E.III.05 shows the planned structure of CETEMM Euvaldo Lodi's Y-CIM-Lab. In order to work out the concept, each cell is modeled by using the ARIS architecture (see *Scheer, ARIS 1992*).

IV. CIM in the Peoples Republic of China: The Chinese 863-program - an introduction

(Prof. Dr. August-Wilhelm Scheer, Mag. Wirtsch.-Ing. Rong Chen, Institut fuer Wirtschaftsinformatik (IWi), University of Saarland)

a. Overview

With the beginning of the 80`s Chinese scientists have started to intensively follow up on worldwide changes taking place in the area of CIM- (Computer Integrated Manufacturing) technologies. The knowledge they gained quickly led to comprehensive discussions about the state of development in *China* and the possibilities of technological exchanges with more industrialized nations. In 1987 these efforts mainly contributed to the establishment of a research and development program called the *"863-Program"*. With a relatively open-

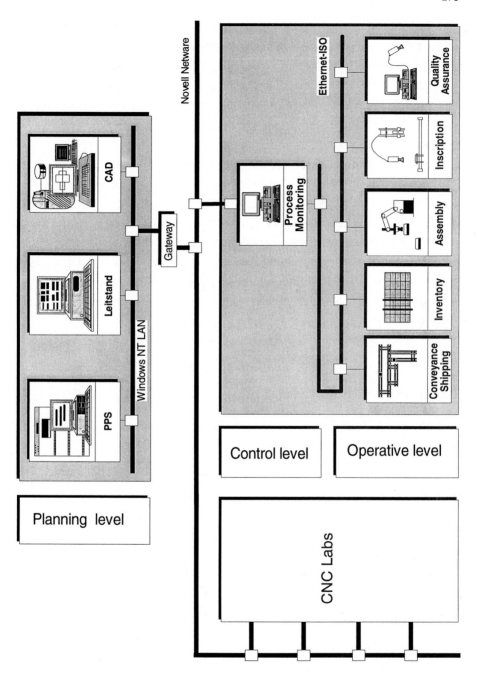

Fig. E.III.05: Planned structure of Y-CIM-Lab at CETEMM Euvaldo Lodi, Rio de Janeiro

handed financial support from the Chinese government researchers were able to substantially accelerate the development of CIM-technologies.

The seven member expert group responsible for all topics concerning CIM within the "863-Program" (*863/CIMS*) had to deal with a general lack of financial resources as well as with an underdeveloped technical situation in the whole area of automation. They put down the general principles in concern of limited objectives, top-down planning and bottom-up implementation. According to these financial support concentrates mainly on aspects of integrating informations. Thereby a framework for the development of further CIM technologies on an equipment based stage was build. The strategic goals until the year 2000 are:

- ❏ to build up test-bed sites in different industries and to introduce CIM according to these industries in different and adjusted ways; the companies will then function as reference to other related ones,
- ❏ to build up sophisticated research labs for CIM technologies and
- ❏ to transform talented engineers and scientists into well educated and hierarchically structured CIM-experts.

b. The General Framework of 863/CIMS

To achieve the projected goals as soon as possible a general framework for 863/CIMS was developed. This framework is shown in Fig. E.IV.01.

The area of CIM is structured into ten different specialized subjects according to which all scientific projects need to be classified. Both, the research and the engineering environment provide the necessary project support. The research environment includes the states CIMS Engineering Research Center (*CIMS-ERC*) and seven other research labs, each concentrating on one or more of the specific subjects (see also Fig. E.IV.02). Furthermore the CIMS-ERC is responsible for the integration of all scientific results into their CIM-system. All of the research labs are located at top Chinese universities and official research centers - their technical equipment and personnel is well above average.

The engineering environment is currently composed of nine enterprises - each a leader within its industry. Their technical as well as economic development is far above average, thus they will serve as reference sites for the 863/CIMS project. The major portion of these companies is in the tool-manufacturing, construction, automobile or aerospace industry.

With Chinas economic reforms in progress the enterprises are more and more confronted with fierce competition. More that ever, they want to take advantage of modern and sophisticated technologies. Even though the 863/CIMS reference sites are supported by

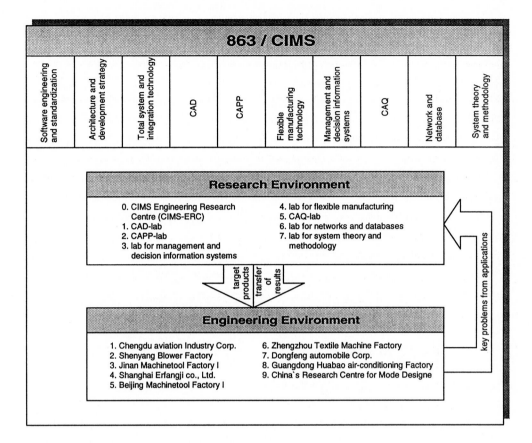

Fig. E.IV.01: General Framework of 863/CIMS

Research Center	Location	Specific Subjects
CIMS-Engineering Research Center (CIMS-ERC)	Tsinghua University, Beijing	CIMS-integration technologies; CIMS-technical measuring CIM-qualification
CAD-lab	Beijing University of Aeronautics and Astronautics, Beijing	CIMS-oriented CAD / CAM / CAE
CAPP-lab	Shanghai Jiao Tong University, Shanghai	CIMS-oriented CAPP
lab for management and decision information systems	Tsinghua University, Beijing	CIMS-oriented management and decision making
lab for flexible manufacturing	Beijing Institute of machinetools, Beijing	CIMS-oriented EMC / FMC / FME
CAQ-lab	Xi´an Jiao Tong University, Xi´an	CIMS-oriented CAQ
lab for networks and databases	Southeast University, Nanjing	CIMS-oriented network and database technologies
lab for system theory and methodology	Shengyang Institute of Automation, Shengyang	CIMS-oriented theory, simulation and artificial intelligence

Fig. E.IV.02: Structure of the 863/CIMS research environment

Source: *Li, B.: The 863/CIMS program 1993, p. 121*

governmental funding, they have invested US-$ 50 mill. in 1992 alone to import the necessary computers, networks, databases, CAD/CAM-systems, MIS, FFS and CNC-machines (see *Wu, C.: The Progress of Chinese Enterprises CIMS Projects 1993, p. 251*)

The "target-products" and the transfer of results both determine the interface between the research and engineering environment. Furthermore the target-products are differentiated into short-term and mid-term products. The first are prototypes which can be applied to industrial use within a years time and can be developed into saleable products from there on. The latter products need further efforts in research and development before they can be applied to industrial use. The transfer of results mainly comprises of a know-how-transfer in seminars, workshops and classes. The scientists involved in 863/CIMS also stay at the reference sites for longer periods of time. They consequently guide projects at all of the reference site locations beginning with conceptual planning until the final implementation. This way dozens of scientists and engineers from the research environment cooperate with the same number of specialists at each of the different reference sites.

c. The Progress of 863/CIMS

The following description is based on the development of CIMS-ERC, because its standard is more or less the highest in China.

The build-up of CIMS-ERC was started in March of 1988 as one of the most important projects within the 863-program. The main goal was to provide a sophisticated environment for research and measurement that Chinese as well as foreign scientists could use to find their ways and means of introducing CIM to Chinese enterprises. Thereby the technological transfer from the research to the engineering community was to be advanced. After a one year planning stage of the overall concept the actual build-up has taken another three and a half years until completion. At the end of 1992 the "demonstration-system" was up and running to its fullest extend. More than 250 scientists and engineers from six different universities and five different official research centers have participated in the build-up. The financial funds amounted up to approximately 27 mill. Yuan - an equivalent of US-$ 5 million. The CIMS/ERC consists of two labs: the information system lab and the manufacturing system lab. Within the information system lab there are more than 30 heterogeneous computers, which are connected to a local area network (LAN) and host a variety of applications including CAD/CAM, distributed database, system simulation, hierarchical control, management and decision information systems. The manufacturing system lab consists of three machining centers, automated guide vehicles (AGV), robots, high-bay stores, three-dimensional measuring machines, tool presenters

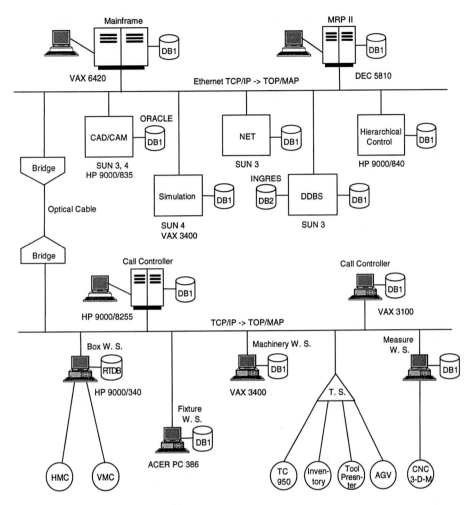

Fig. E.IV.03: CIMS/ERC system architecture
from: *Fan, Y., Wu, C.: The CIMS-ERC 1991, p. 37*

and management systems. The task of planning and control of shop-floor equipment is handled in a computer room with its own LAN. In Fig. E.IV.03 the CIMS/ERC architecture is illustrated.

According to the evaluation of Chinese experts the CIMS/ERC has already reached the advanced level of famous CIMS centers in the world, which are build-up and running by the end of the 80`s.

Most of the nine reference sites have by now finished their conceptual system planning and are on the implementation stage right now. First results and benefits can be expected in the second half of 1994.

Besides the 863/CIMS project there are regional or industry related efforts and initiatives to research and apply CIM-technologies. More than 20 Chinese universities have their own CIM research centers for example. Still the extend and state of development of these institutions is differing significantly.

d. Prospect

Along with the development of CIM in China the focal point has drastically changed from an emphasis on technical problems to an emphasis on economic questions. Amongst other things, this can be seen at the high interest of Chinese scholars in simultaneous engineering and lean production. They are eager to link the latest scientific findings in these areas with their own research.

More than 100 small and medium sized enterprises from different industries will be selected to implement CIM-technologies into their production processes over the next few years. These enterprises can rely on strong governmental funding and intensive scientific support. Their chances to get excellent results in a short period of time will be by far better than today.

In research as well as in industry the progress is differing substantially from one site to the next. Few enterprises have come up with their own CIM-technologies due to financial and professional reasons. Most Chinese managers are far from having the appropriate knowledge and training concerning CIM. A comprehensive and successful development of CIM in China will therefor take a lot of time and very strong efforts by all people involved.

F. References

Ahlers, J., Gröner, L., Mattheis, P.: *Konstruktionsbegleitende Kalkulation im Rahmen des CAD 1986*
Kalkulation im Rahmen des Computer Aided Design, Studie des Instituts für Wirtschaftsinformatik für die Firma Siemens AG, München 1986.

Anselstetter, R.: *Nutzeffekte der Datenverarbeitung 1986*
Betriebswirtschaftliche Nutzeffekte der Datenverarbeitung, 2nd ed., Berlin, Heidelberg, New York, Tokyo 1986.

Ausschuß für Wirtschaftliche Fertigung (AWF, ed.): *Software für die Fertigung 1988*
Software für die Fertigung, Eschborn 1988.

Ausschuß für Wirtschaftliche Fertigung (AWF, ed.): *Expertensysteme in der Praxis 1988*
Expertensysteme in der betrieblichen Praxis - Ergebnisse des AWF/VDI-Arbeitskreises, Eschborn 1988.

Becker, J.: *EDV-System zur Materialflußsteuerung 1987*
Architektur eines EDV-Systems zur Materialflußsteuerung, Berlin, Heidelberg, New York,

Blum, Engelkamp, Porten: *Kommunikationsnetze für CIM 1988*
Kommunikationsnetze für CIM, in: Schulungsunterlagen des CIM-Technologie-Transfer-Zentrums Stuttgart, Querschnittsthema 6 "Netze und Kommunikationstechnik", 1988.

Böhm, E.: *Konfiguration komplexer Produkte 1986*
Anwendung von Expertensystemen zum Konfigurieren und Anpassen komplexer Produkte, in: Warnecke, H. J., Bullinger, H.-J. (eds.), 18. Arbeitstagung des IPA (Institut für Produktionstechnik und Automatisierung), Berlin, Heidelberg 1986.

Brockhoff, K., Picot, A., Urban, C. (eds.): *Zeitmanagement 1988*
Zeitmanagement in Forschung und Entwicklung, in: ZfbF, (1988) special no. 23.

Bullinger, H.-J., Warnecke, H. J., Lentes, H.-P.: *Factory of the Future 1985*
Toward the Factory of the Future, Opening Adress, in: Bullinger, H.-J., Warnecke, H. J. (eds.), Kornwachs K. (assistant ed.), Toward the Factory of the Future, pp. XXIX - LIV, Berlin, Heidelberg, New York, Tokyo 1985.

Commission of the European Communities (ed.): *ESPRIT Workprogramme 1989*
1989 ESPRIT Workprogramme. Brussels 1989.

Commission of the European Communities (ed.): *The Project Synopses CIM 1989*
ESPRIT - The Project Synopses Computer Integrated Manufacturing, vol. 6, Brussels 1989.

Diedenhoven, H.: *Informationsgehalt von CAD-Daten 1985*
Für die NC-Fertigung nutzbarer Informationsgehalt von CAD-Daten, in: CAE-Journal, (1985) no. 5, pp. 58 - 65.

Dittrich, K. R., et al.: *Datenbankunterstützung für den ingenieurwissenschaftlichen Entwurf 1985*
Datenbankunterstützung für den ingenieurwissenschaftlichen Entwurf, in: Informatik-Spektrum, 8 (1985) no. 3, pp. 113 - 125.

Ehrlenspiel, K.: *Kostengünstig Konstruieren 1985*
Kostengünstig Konstruieren, Berlin, Heidelberg, New York, Tokyo 1985.

Eidenmüller, B.: *Produktion als Wettbewerbsfaktor 1989*
Die Produktion als Wettbewerbsfaktor, Herausforderungen an das Produktionsmanagement, Köln 1989.

Encarnaçao, J., et al. (eds.): *CAD-Handbuch 1984*
CAD-Handbuch, Auswahl und Einführung von CAD-Systemen, Berlin, Heidelberg, New York, Tokyo 1984.

ESPRIT Consortium AMICE (ed.): *CIM-OSA 1989*
CIM-OSA, ESPRIT Project No. 688: Reference Architecture Specification, Brussels 1989.

Eversheim, W.: *Simultaneous Engineering 1989*

Simultaneous Engineering - eine organisatorische Chance!, in: VDI Berichte, 758 (Simultaneous Engineering: Tagung, Frankfurt, 18. und 19. April 1989), Düsseldorf 1989.

Fan, Y., Wu, C.: *The CIMS-ERC 1991*

The Architecture and Information Integration of the State CIMS Engineering Research Center, 1991, Proceedings of the International Conference on Computer Integrated Manufacturing, Singapore, Oct. 1991, pp. 34 - 37.

Fischer, W.: *Datenbank-Management in CAD/CAM-Systemen*

Datenbank-Management in CAD/CAM-Systemen: Anforderungen an CAD/CAM Datenbanken - Derzeitiges Marktangebot - Empfehlungen für effizientes DB-Management - Entwicklungstendenzen, Messerschmitt-Bölkow-Blohm GmbH, Ottobrunn.

Geitner, U. W. (ed.): *CIM-Handbuch 1987*

CIM-Handbuch, Wiesbaden 1987.

Gora, W.: *MAP 1986*

MAP, in: Informatik-Spektrum, 9 (1986) no. 1, pp. 40 - 42.

Grabowski, H., Glatz, R.: *Schnittstellen 1986*

Schnittstellen zum Austausch produktdefinierender Daten, Auf dem Weg zum internationalen Standard STEP, in: VDI-Z, 128 (1986) no. 18, pp. 333 - 343.

Gröner, L., Roth, L.: *Konzeption eines CIM-Managers 1986*

Konzeption eines CIM-Managers, in: Logistik Heute, (1986) no. 10, pp. II - VI.

Gunn, T. G.: *Konstruktion und Fertigung 1982*

Konstruktion und Fertigung, in: Spektrum der Wissenschaft, November 1982, pp. 77 - 98.

Haberstroh, G., Nölscher, N.: *Im Netz von CIM 1989*

Im Netz von CIM, in: UNIX-Magazin, (1989) no. 9, p. 88.

Hackstein, R.: *PPS 1984*

Produktionsplanung und -steuerung (PPS), Ein Handbuch für die Betriebspraxis, Düsseldorf 1984.

Hackstein, R.: CIM-Begriffe sind verwirrende Schlagwörter 1985
CIM-Begriffe sind verwirrende Schlagwörter - Die AWF-Empfehlung schafft Ordnung, in: AWF-Ausschuß für Wirtschaftliche Fertigung e.V. (ed.), PPS 85, Proceedings, Böblingen 1985.

Hahn, D., Laßmann, G. (eds.): *Produktionswirtschaft 1989*
Produktionswirtschaft - Controlling industrieller Produktion, vol. 2: Produktionsprozesse - Grundlegung zur Produktionsprozeßplanung, -steuerung und -kontrolle und Beispiele aus der Wirtschaftspraxis, Heidelberg 1989.

Harmon, P., King, D.: *Expertensysteme in der Praxis 1988*
Expertensysteme in der Praxis: Perspektiven, Werkzeuge, Erfahrungen, München 1988.

Harrington, J.: Computer Integrated Manufacturing 1979
Computer Integrated Manufacturing, Reprint, Malabar (Florida) 1979.

Hedrich, P., et al.: *Flexibilität in der Fertigungstechnik 1983*
Flexibilität in der Fertigungstechnik durch Computereinsatz, München 1983.

Hellwig, H.-E., Hellwig, U.: *CIM-Konzepte 1986*
CIM-Konzepte und CIM-Bausteine, in: VDI-Z, 128 (1986) no. 18, pp. 691 - 703.

Hewlett Packard (ed.): *HP CIM-Server 1988*
HP CIM-Server, 1988.

Hübel, C.: *Datenbankorientierter 3-D-Bauteilmodellierer 1985*
Ein datenbankorientierter 3-D-Bauteilmodellierer als Anwendung eines Nicht-Standard-Datenbanksystems - Ansätze zur quantitativen Systemanalyse, Technischer Bericht Fachbereich Informatik, Universität Kaiserslautern, 1985.

IDS Prof. Scheer GmbH (ed.): *Leitstand 1992*
Dezentrale Fertigungssteuerung: der Intelligente Leitstand FI-2. Systembeschreibung, Saarbrücken 1990. IDS Prof. Scheer GmbH (ed.): Der intelligente Leitstand FI-2 - Benutzerhandbuch, Band 1, Version 2.70, Saarbrücken 1991. IDS Prof. Scheer GmbH (ed.): Der intelligente Leitstand FI-2 - Benutzerhandbuch Band 2, Version 2.70, Saarbrücken 1991.

Ives, B, Learmonth, G. P.: *The Information System as a Competitive Weapon 1984*
The Information System as a Competitive Weapon, in: Communications of the
ACM, 27 (1984) no. 12, pp. 1193 - 1201.

Janetzky, D., Schwarz, K.: *Das MAP-Projekt 1985*
Das MAP-Projekt, Technik, Stand und Aktivitäten, in: VDI/VDE-Gesellschaft
Meß- und Regelungstechnik (GMR, ed.), Proceedings zum Workshop Technische
Kommunikation in der Automatisierungstechnik, GMR-Bericht 8, Düsseldorf 1985.

Jost, W.: *Werkzeugunterstützung 1993*
Werkzeugunterstützung in der DV-Beratung, in: Information Management, no. 8
(1993)1, pp. 10 - 19.

Kauffels, F.-J.: *Klassifizierung der lokalen Netze 1986*
Klassifizierung der lokalen Netze, in: Neumeier, H. (ed.), State of the Art: Lokale
Netze, (1986) no. 2, München 1986, pp. 5 - 13.

Kazmeier, E.: *Belastungssituation im Rahmen eines PPS-Systems 1984*
Berücksichtigung der Belastungssituation im Rahmen eines neuen PPS-Systems auf
der Basis einer dialogorientierten Ablaufplanung, in: Institut für Fertigungsanlagen
der Universität Hannover (IFA, ed.), Statistisch orientierte Fertigungssteuerung,
Hannover 1984.

Kernler, H.: *Einsatzspektrum des PC's für PPS-Aufgaben 1985*
Einsatzspektrum des PC's für PPS-Aufgaben - Ergebnisse einer Pilotarbeit; in:
AWF-Ausschuß für Wirtschaftliche Fertigung e.V. (ed.), PPS 85, Proceedings, Böb-
lingen 1985.

Kief, H. B.: *NC Handbuch 1984*
NC Handbuch, Michelstadt, Stockheim 1984.

Kilger, W.: *Einführung in die Kostenrechnung 1987*
Einführung in die Kostenrechnung, 3rd ed., Wiesbaden 1987.

Kochan, A., Cowan, D.: *Implementing CIM 1986*
Implementing CIM - Computer Integrated Manufacturing, Bedford and Berlin, Heidelberg, New York, Tokyo 1986.

König, W., Hennicke, L.: *PROPEX 1987*

Das Produktionsplanungs-Expertensystem PROPEX, Entwicklung und Einsatzperspektiven, in: Wildemann, H. (ed.), Expertensysteme in der Produktion, München 1987.

Krallmann, H: *Anwendungen in CIM 1986*

Anwendungen in CIM, in: Schnupp, P. (ed.), State of the Art: Expertensysteme, (1986) no. 1, München 1986, pp. 73 - 78.

Krallmann, H.: *EES 1986*

EES - das Expertensystem für den Einkauf, in: Betriebswirtschaftliche Forschung und Praxis (BFuP), 6 (1986), pp. 565 - 583.

Krallmann, H.: *Expertensysteme für CIM 1986*

Expertensysteme für die computerintegrierte Fertigung, in: Warnecke, H. J., Bullinger, H. J. (eds.), 18. Arbeitstagung des IPA (Institut für Produktionstechnik und Automatisierung), Berlin, Heidelberg 1986.

Krallmann, H.: *Expertensysteme in PPS 1987*

Expertensysteme in der Produktionsplanung und -steuerung, in: CIM-Management, 3 (1987) no. 4, pp. 60 - 69.

Krallmann, H.: *PPS-Expertensysteme 1987*

PPS-Expertensysteme: Aufbau, Einsatzmöglichkeiten und Wirtschaftlichkeit, in: Geitner, U. W. (ed.), CIM-Handbuch, Wiesbaden 1987, pp. 287 - 321.

Kreisfeld, P.: *Kostenbestimmung mit CAD-Systemen 1985*

Kostenbestimmung mit CAD-Systemen für Rotationsteile, in: Spur, G. (ed.), Produktionstechnik - Berlin, Forschungsberichte für die Praxis, München, Wien 1985.

Lechner, K.-O., Schuy, K.: *Computer Integrated Manufacturing 1986*

Computer Integrated Manufacturing oder das Ende der Kreativität, in: Der stumme Dialog, Stuttgart 1986, pp. 211 - 231.

Li, B.: *The 863/CIMS program 1993*

Overview of the 863/CIMS program, China Computerworld, 12/01/1993, p. 121 (in Chinese).

Lingnau, H. E.: *Realisierung eines CIM-Konzepts 1985*
Vorgehensweise und Möglichkeiten zur Realisierung eines CIM-Konzepts unter Berücksichtigung vorhandener EDV-Instrumente, in: AWF-Ausschuß für Wirtschaftliche Fertigung e.V. (ed.), PPS 85, Proceedings, Böblingen 1985.

McFarlan, F. W.: *Information Technology Changes the Way You Compete 1984*
Information Technology Changes the Way You Compete, in: Harvard Business Review, 62 (1984) no. 3, pp. 98 - 103.

Mertens, P.: *Zwischenbetriebliche Integration 1985*
Zwischenbetriebliche Integration der EDV, in: Informatik-Spektrum, 8 (1985) no. 2, pp. 81 - 90.

Mertens, P.: *Wissensbasierte Systeme in der PPS 1988*
Wissensbasierte Systeme in der Produktionsplanung und -steuerung - Eine Bestandsaufnahme, in: Information Management (IM), 3 (1988) no. 4, pp. 14 - 22.

Mertens, P.: *Verbindung von verteilter PPS und verteilten ES 1989*
Verbindung von verteilter Produktionsplanung und -steuerung und verteilten Expertensystemen, in: Information Management (IM), 4 (1989) no. 1, pp. 6 - 11.

Mertens, P., Allgeyer, K., Däs, H.: *Betriebliche Expertensysteme 1986*
Betriebliche Expertensysteme in deutschsprachlichen Ländern - Versuch einer Bestandsaufnahme, Arbeitsberichte des Instituts für mathematische Maschinen und Datenverarbeitung (Informatik), Friedrich-Alexander-Universität, Erlangen-Nürnberg, Erlangen 1986.

Mertens, P., Borkowski, V., Geis, W.: *Expertensystem-Anwendungen 1988*
Betriebliche Expertensystem-Anwendungen, Eine Materialsammlung, Berlin, Heidelberg, New York, London, Paris, Tokyo 1988.

Neckermann, R.: *Das Netz von morgen 1985*
Das Netz von morgen wird heute schon gestrickt, Sonderdruck aus Produktion, 25 (1985).

Nüttgens, M.; Scheer, A.-W.: *CIM-Qualifizierungskonzept für KMU 1991*,
CIM-Qualifizierungskonzept für Klein- und Mittelunternehmungen (KMU), published in Institut Technik und Bildung (editor): CIM Qualifizierung in Europa, Proceedings of the same named conference on 19.09.91 in Bremen, pp. 229-240

Nüttgens M.; Keller, G.; Scheer A.-W.: Information Controlling 1992
Informationsmodell für ein integriertes Informations-Controlling. In: HMD no. 29 (1992) 166, pp. 122-126.

Nüttgens, M.; Scheer, A.-W.: ARIS-Navigator 1993
ARIS-Navigator. In : Information Management, no. 8 (1993)1, pp. 20-26.

Opitz, H.: Klassifizierungssystem 1966
Werkstückbeschreibendes Klassifizierungssystem, Essen 1966.

Porter, M. E.: Creating and Sustaining Superior Performance 1985
Competitive Advantage: Creating and Sustaining Superior Performance, The Free Press, New York 1985.

Puente, E., MacConaill, P. (eds.): Computer Integrated Manufacturing 1988
Computer Integrated Manufacturing, Proceedings of the 4th CIM Europe Conference, Bedford and Berlin, Heidelberg, New York, London, Paris, Tokyo 1988.

Puppe, F.: Expertensysteme 1986
Expertensysteme, in: Informatik-Spektrum, 9 (1986) no. 1, pp. 1 - 13.

Puppe, F.: Einführung in ES 1988
Einführung in Expertensysteme, Berlin, Heidelberg, New York 1988.

Ránky, P. G.: Computer Integrated Manufacturing 1986
Computer Integrated Manufacturing, An Introduction with Case Studies, Englewoods Cliffs, New Jersey, London, Mexico, New Delhi, Rio de Janeiro, Singapore, Sydney, Tokyo, Toronto 1986.

Rausch, W., de Marne, K.-D.: VDA-Flächenschnittstelle 1985
Datenaustausch über die VDA-Flächenschnittstelle mit CAD/CAM-System STRIM 100, in: CAD/CAM, (1985) no. 6, pp. 54 - 61, und CAD/CAM, (1986) no. 1, pp. 95 - 102.

Reitzle, W.: Industrieroboter 1984
Industrieroboter, München 1984.

Remboldt, U., Dillmann, R. (eds.): *Computer-Aided Design and Manufacturing 1986*
Computer-Aided Design and Manufacturing, Methods and Tools, 2nd ed., Berlin, Heidelberg, New York, London, Paris, Tokyo 1986.

Rockart, J. F.: *Chief Executives Define their Own Data Needs 1979*
Chief Executives Define their Own Data Needs, in: Harvard Business Review, 57 (1979) no. 2, pp. 81 - 93.

Rolle, G.: *ES für PC 1988*
Expertensysteme für Personalcomputer, Würzburg 1988.

Rome, E., Uthmann, T., Diederich, J.: *KI-Workstation 1988*
KI-Workstations, Überblick - Marktsituation - Entwicklungstrends, Bonn 1988.

Savory, S.: *Künstliche Intelligenz 1985*
Künstliche Intelligenz und Expertensysteme, Ein Forschungsbericht der Nixdorf Computer AG, 2nd ed., München 1985.

Schäfer, H.: *Technische Grundlagen der lokalen Netze 1986*
Technische Grundlagen der lokalen Netze, in: Neumeier, H. (ed.), State of the Art: Lokale Netze, (1986) no. 2, München 1986, pp. 14 - 23.

Scheer, A.-W.: *Instandhaltungspolitik 1974*
Instandhaltungspolitik, Wiesbaden 1974.

Scheer, A.-W.: *Produktionsplanung auf der Grundlage einer Datenbank 1976*
Produktionsplanung auf der Grundlage einer Datenbank des Fertigungsbereichs, München, Wien 1976.

Scheer, A.-W.: *Computergestützte PPS 1983*
Stand und Trends der computergestützten Produktionsplanung und -steuerung (PPS) in der Bundesrepublik Deutschland, in: Zeitschrift für Betriebswirtschaft, 53 (1983), pp. 138 - 155.

Scheer, A.-W.: *Aufgabenverteilung Mikro-Mainframe 1985*
Kriterien für die Aufgabenverteilung in Mikro-Mainframe-Anwendungssystemen, in: Scheer, A.-W. (ed.), Veröffentlichungen des Instituts für Wirtschaftsinformatik, no. 48, Saarbrücken 1985.

Scheer, A.-W.: *Wirtschaftlichkeitsfaktoren 1985*
Wirtschaftlichkeitsfaktoren EDV-orientierter betriebswirtschaftlicher Problemlösungen, in: Scheer, A.-W. (ed.), Veröffentlichungen des Instituts für Wirtschaftsinformatik, no. 49, Saarbrücken 1985.

Scheer, A.-W.: *Konstruktionsbegleitende Kalkulation in CIM 1985*
Konstruktionsbegleitende Kalkulation in CIM-Systemen, in: Scheer, A.-W. (ed.), Veröffentlichungen des Instituts für Wirtschaftsinformatik, no. 50, Saarbrücken 1985.

Scheer, A.-W.: *Strategie zur Entwicklung eines CIM-Konzeptes 1986*
Strategie zur Entwicklung eines CIM-Konzeptes - Organisatorische Entscheidungen bei der CIM-Implementierung, in: Scheer, A.-W. (ed.), Veröffentlichungen des Instituts für Wirtschaftsinformatik, no. 51, Saarbrücken 1986.

Scheer, A.-W.: *Entscheidungsunterstützungssysteme 1986*
Entscheidungsunterstützungssysteme, in: FORUM, (1986) no. 5, pp. 11 - 17.

Scheer, A.-W.: *Neue PPS-Architekturen 1986*
Neue Architektur für EDV-Systeme zur Produktionsplanung und -steuerung, in: Scheer, A.-W. (ed.) Veröffentlichungen des Instituts für Wirtschaftsinformatik, no. 53, Saarbrücken 1986.

Scheer, A.-W.: *Production Systems 1986*
Production Control and Information Systems, in: Remboldt, U., Dillmann, R. (eds.), Computer-Aided Design and Manufacturing, Methods and Tools, 2nd ed., Berlin, Heidelberg, New York, London, Paris, Tokyo 1986.

Scheer, A.-W. (ed.): *Betriebliche Expertensysteme I 1988*
Betriebliche Expertensysteme I (Einsatz von Expertensystemen in der Betriebswirtschaft - Eine Bestandsaufnahme), in: Jacob, H., et al. (eds.), Schriften zur Unternehmensführung (SzU), vol. 36, Wiesbaden 1988.

Scheer, A.-W.: *CIM in den USA 1988*
CIM in den USA - Stand der Forschung, Entwicklung und Anwendung, in: Scheer, A.-W. (ed.), Veröffentlichungen des Instituts für Wirtschaftsinformatik, no. 58, Saarbrücken 1988.

Scheer, A.-W.: *CIM in Brasilien 1991*

CIM in Brasilien, in Scheer, A.-W.:(ed.), Information Management, 6 th annual set, 1991, No.2, pp. 64 - 69.

Scheer, A.-W. (ed.): *Betriebliche Expertensysteme II 1989*

Betriebliche Expertensysteme II (Einsatz von Expertensystem-Prototypen in betriebswirtschaftlichen Funktionsbereichen), in: Jacob, H., et al. (eds.), Schriften zur Unternehmensführung (SzU), vol. 40, Wiesbaden 1989.

Scheer, A.-W.: *EDV-orientierte BWL 1990*

EDV-orientierte Betriebswirtschaftslehre - Grundlagen für ein effizientes Informationsmanagement, 4th ed., Berlin, Heidelberg, New York, London, Paris, Tokyo, Hong Kong, Barcelona 1990.

Scheer, A.-W.: *Wirtschaftsinformatik 1990*

Wirtschaftsinformatik - Informationssysteme im Industriebetrieb, 3rd ed., Berlin, Heidelberg, New York, London, Paris, Tokyo, Hong Kong, Barcelona 1990.

Scheer, A.-W.: *Enterprise-Wide Data Modelling 1989*

Enterprise-Wide Data Modelling - Information Systems in Industry, Berlin, Heidelberg, New York, London, Paris, Tokyo, Hong Kong 1989.

Scheer, A.-W.: *Communication Technology 1991*

Konsequenzen für die Betriebswirtschaftslehre aus der Entwicklung der Informations- und Kommunikationstechnologien. Veröffentlichung des Instituts für Wirtschaftsinformatik, Heft 79, Saarbrücken 1991.

Scheer, A.-W.: *Principles of efficient information management 1991*

Principles of efficient information management, Berlin, Heidelberg, New York, London, Paris, Tokyo, Hong Kong, Barcelona 1991.

Scheer, A.-W.: *ARIS 1992*

Architecture for Integrated Information Systems (ARIS) - Foundations of Enterprise Modelling. Berlin 1992.

Scheer, A.-W., Becker, J., Bock, M.: *Expertensystem zur konstruktionsbegleitenden Kalkulation 1988*
Ein Expertensystem zur konstruktionsbegleitenden Kalkulation, in: Gollan, B., Paul, W. J., Schmitt, A. (eds.), Innovative Informations-Infrastrukturen, Berlin, Heidelberg, New York, London, Paris, Tokyo 1988.

Scheer, A.-W., Kraemer, W.: *Konzeption und Realisierung I 1989*
Konzeption und Realisierung eines Expertenunterstützungssystems im Controlling, in: Scheer, A.-W. (ed.), Veröffentlichungen des Instituts für Wirtschaftsinformatik, no. 60, Saarbrücken 1989.

Scheer, A.-W., Kraemer, W.: *Konzeption und Realisierung II 1989*
Konzeption und Realisierung eines Expertenunterstützungssystems im Controlling, in: Kurbel, K., Mertens, P., Scheer, A.-W. (eds.), Interaktive betriebswirtschaftliche Informations- und Steuerungssysteme, Studien zur Wirtschaftsinformatik, vol. 3, Berlin, New York 1989, pp. 157 - 184.

Scheer, A.-W., Steinmann, D.: *Einführung in ES 1988*
Einführung in den Themenbereich Expertensysteme, in: Jacob, H., et al. (eds.), Betriebliche Expertensysteme I, Schriften zur Unternehmensführung (SzU), vol. 36, Wiesbaden 1988, pp. 5 - 27.

Scheer, A.-W., Steinmann, D.: *WBS in der PPS I 1989*
Wissensbasierte Systeme in der Produktionsplanung und -steuerung (PPS) in Computer Integrated Manufacturing (CIM)-Systemen, in: CAD/CAM-Report, 8 (1989) no. 4, pp. 109 - 115.

Scheer, A.-W., Steinmann, D.: *WBS in der PPS II 1989*
Wissensbasierte Systeme in der Produktionsplanung und -steuerung (PPS) in Computer Integrated Manufacturing (CIM)-Systemen, in: CAD/CAM-Report, 8 (1989) no. 5, pp. 52 - 65.

Schliep, W.: ES zur Fehlerdiagnose an fahrerlosen Transportsystemen 1988
Expertensystem zur Fehlerdiagnose an fahrerlosen Transportsystemen, in: Ausschuß für Wirtschaftliche Fertigung (AWF, ed.), Expertensysteme in der betrieblichen Praxis - Ergebnisse des AWF/VDI-Arbeitskreises, Eschborn 1988.

Schneider, J.: *Datenübertragung 1986*

Datenübertragung von VDA-Datensätzen mit dem FTP, in: DATACOM, 5 (1986), pp. 64 - 68.

Schnupp, P.: *Rechnernetze 1982*

Rechnernetze - Entwurf und Realisierung, 2nd ed., Berlin, New York 1982.

Schnupp, P., Leibrandt, U.: *Expertensysteme 1988*

Expertensysteme, 2nd ed., Berlin, Heidelberg, Tokyo 1988.

Scholz, B.: *CIM-Schnittstellen 1988*

CIM-Schnittstellen, Konzepte, Standards und Probleme der Verknüpfung von Systemkomponenten in der rechnerintegrierten Produktion, München, Wien 1988.

Schümmer, M.: *MMS/RS-511 1988*

Manufacturing Message Specification MMS/RS-511, in: Informatik-Spektrum, 11 (1988) no. 4, pp. 209 - 211.

Schwindt, P.: *CAD-Austausch 1986*

CAD-Datenaustausch aus der Sicht eines mittelständischen Automobilzulieferers, in: Stuttgarter Messe- und Kongreß-GmbH (ed.), CAT'86 (Computer Aided Technologies in Manufacturing), Leinfelden, Echterdingen 1986, pp. 38 - 40.

Scown, S. J.: *Artificial Intelligence 1985*

The Artificial Intelligence Experience: An Introduction, Digital Equipment Corporation 1985.

Simon, T.: *Kommunikation in der automatisierten Fertigung 1986*

Kommunikation in der automatisierten Fertigung, in: Computer Magazin, (1986) no. 6, pp. 38 - 42.

Smith, G.: *OPT-Realisierung 1985*

OPT-Realisierung - Einsatzerfahrungen mit der OPT-Software, in: GF+M (ed.), Produktionsmanagement - heute realisiert, Jahrestagung 1985, p. 67 ff.

Sock, E., Nagel J.: *CAD/CAM-Integration 1986*

Acht Thesen zur Verwirklichung der CAD/CAM-Integration, in: CIM Management, 2 (1986) no. 1, pp. 18 - 25.

Sorgatz, U., Hochfeld, H.-J.: *Austausch produktdefinierender Daten 1985*
Austausch produktdefinierender Daten im Anwendungsgebiet der Karosseriekon-struktion, in: Informatik-Spektrum, 8 (1985) no. 6, pp. 305 - 311.

Spur, G., Krause F.-L.: *CAD-Technik 1984*
CAD-Technik, München, Wien 1984.

Steinacker, I.: *Expertensystem als Bindeglied zwischen CAD und CAM 1985*
Ein Expertensystem als Bindeglied zwischen CAD und CAM, in: Trost, H., Retti, J. (eds.), Informatik Fachberichte, vol. 106, Subreihe Künstliche Intelligenz, Öster-reichische Artifical Intelligence Tagung (Wien), Berlin, Heidelberg, New York, Tokyo 1985.

Steinmann, D.: *ES in der PPS unter CIM-Aspekten 1987*
Expertensysteme (ES) in der Produktionsplanung und -steuerung (PPS) unter CIM-Aspekten, in: Scheer, A.-W., Veröffentlichungen des Instituts für Wirtschaftsinfor-matik, no. 55, Saarbrücken 1987.

Steinmann, D.: *Standard- und/oder Individualsoftware 1988*
Standard- und/oder Individualsoftware, in: Ausschuß für Wirtschaftliche Fertigung (AWF, ed.), Software für die Fertigung, Eschborn 1988, pp. 242 - 272.

Steinmann, D.: *Konzeption zur Integration von WBS in PPS 1989*
Konzeption zur Integration wissensbasierter Anwendungen in konventionelle Systeme der Produktionsplanung und -steuerung (PPS) im Bereich der Fertigungs-steuerung, in: Jacob, H., et al. (eds.), Betriebliche Expertensysteme II, Schriften zur Unternehmensführung (SzU), vol. 40, Wiesbaden 1989, pp. 83 - 122.

Suppan-Borowka, J.: *Anforderungen an MAP 1986*
MAP unter der Lupe - Anforderungen an MAP, in: Technische Rundschau, 78 (1986), pp. 170 - 175.

Suppan-Borowka, J., Simon T.: *MAP in der automatisierten Fertigung 1986*
MAP Datenkommunikation in der automatisierten Fertigung, Pulheim 1986.

Teicholz, E.: *Computer Integrated Manufacturing 1984*
Computer Integrated Manufacturing, in: Datamation, March 1984, pp. 169 - 174.

Trum, P.: *Automatische Generierung von Arbeitsplänen 1986*

Automatische Generierung von Arbeitsplänen, in: Schnupp, P. (ed.), State of the Art: Expertensysteme, (1986) no. 1, München 1986, pp. 69 - 72.

Turksen, I. B.: *Computer Integrated Manufacturing 1988*

Computer Integrated Manufacturing, Current Status and Challenges, Proceedings of the NATO Advanced Study Institute on Computer Integrated Manufacturing: Current Status and Challenges, in: NATO ASI Series, Series F: Computer and Systems Sciences, vol. 49, Berlin, Heidelberg, New York, London, Paris, Tokyo 1988.

Vernadat, F.: *Designing Logical Schemata 1986*

Designing Logical Schemata for Manufacturing Databases, in: Technical Paper MS86-726, Proceedings of the 5th Canadian CAD/CAM and Robotics Conference, pp. MS86-726-1 - MS86-726-9.

Wiendahl, H.-P.: *Verfahren der Fertigungssteuerung 1984*

Grundlagen neuer Verfahren der Fertigungssteuerung, in: Institut für Fabrikanlagen der Universität Hannover (IFA, ed.), Statistisch orientierte Fertigungssteuerung, Hannover 1984, pp. 1 - 19.

Wiendahl, H.-P., Lüssenhop, T.: *Basis eines Expertensystems 1986*

Ein neuartiges Produktionsprozeßmodell als Basis eines Expertensystems für die Fertigungssteuerung, in: Warnecke, H. J., Bullinger, H. J. (eds.), 18. Arbeitstagung des IPA (Institut für Produktionstechnik und Automatisierung), Berlin, Heidelberg 1986.

Wildemann, H. (ed.): *ES in der Produktion 1987*

Expertensysteme in der Produktion, München 1987.

Wildemann, H., et al.: *Flexible Werkstattsteuerung 1984*

Flexible Werkstattsteuerung durch Integration von KANBAN-Prinzipien, in: Wildemann, H. (ed.), Computergestütztes Produktionsmanagement, vol. 2, München 1984.

Wiseman, C.: *Information Systems as Competitive Weapons 1985*

Strategy and Computers: Information Systems as Competitive Weapons, Dow Jones-Irwin, Homewood, Illinois 1985.

Wittemann, N.: *Produktionsplanung mit verdichteten Daten 1985*
> Produktionsplanung mit verdichteten Daten, in: Hansen, H. R., et al. (eds.), Betriebs- und Wirtschaftsinformatik, vol. 14, Berlin, Heidelberg, New York, Tokyo 1985.

Womack, J.P.; Jones, D.T.; Roos, D.T.: *Automobil 1992*
> Die zweite Revolution in der Automobilindustrie, 6. Auflage, Frankfurt 1992.

Wu, C.: *The Progress of Chinese Enterprises CIMS Projects 1993*
> The Progress of Chinese Enterprises CIMS Projects and the Technical Support of the State CIMS Engineering Research Center, Proceedings of the 2nd International Conference on Manufacturing Technology, Hong Kong, Dec. 1993, pp. 250 - 252.

Zeilinger, P.: *Just-in-time und DFÜ bei BMW 1986*
> Just-in-time und DFÜ bei BMW, in: ACTIS GmbH, Just-in-time mit DFÜ-Entwicklung und Stand der Datenfernübertragung in der Automobilindustrie, Proceedings, 29./30. September 1986.

Zelm, M.: *Enterprise Modelling 1989*
> Enterprise Modelling, in: ESPRIT CIM, CIM-Europe Workshop Open Systems Architectures and Communications "Preparing the Enterprise for CIM", Workshop Proceedings, Aachen 1989.

op. cit.: *MS-DOS 5.0 1986*
> MS-DOS 5.0 im Alpha-Test, in: Infowelt, (1986) no. 31/12, p. 3.

G. Index